Occupational Health and Hygiene

Jeremy Stranks
MSc, FCIEH, FIOSH, RSP

PITMAN PUBLISHING
128 Long Acre, London WC2E 9AN

A Division of Pearson Professional Limited

First published in Great Britain, 1995

© Jeremy Stranks 1995

British Library Cataloguing in Publication Data
A CIP catalogue record for this book can be obtained from the British Library.

ISBN 0 273 60908 4

All rights reserved; no part of this publication may be reproduced, stored in a retrieval system, or transmitted in any form or by any means, electronic, mechanical, photocopying, recording, or otherwise without either the prior written permission of the Publishers or a licence permitting restricted copying in the United Kingdom issued by the Copyright Licensing Agency Ltd, 90 Tottenham Court Road, London W1P 9HE. This book may not be lent, resold, hired out or otherwise disposed of by way of trade in any form of binding or cover other than that in which it is published, without the prior consent of the Publishers.

10 9 8 7 6 5 4 3 2 1

Typeset by Northern Phototypesetting Co. Ltd, Bolton
Printed and bound in Great Britain by
Bell & Bain Ltd, Glasgow

The Publishers' policy is to use paper manufactured from sustainable forests.

Contents

...O TECHNICAL COLLEGE

Preface

The number of days lost through occupationally-induced ill health and disease in many organisations outweighs time lost as a result of accidents at work. The direct cost to the UK, in terms of days lost from work and the payment of various health-related benefits, is well-known but the indirect costs are less obvious. There is a need, therefore, for organisations to devote more time and resources to the examination of the causes of occupational ill health and the introduction of strategies to prevent or control the exposures which create these losses.

The last decade has seen the introduction of subordinate legislation principally directed at protecting the health of people at work. In particular, the Control of Substances Hazardous to Health (COSHH) Regulations 1988 and 1994 brought in a new regime of worker health protection requiring employers to undertake health risk assessments, monitor exposure to hazardous substances and provide health surveillance where appropriate. More recent European-driven legislation has further reinforced the duties of employers towards their employees and others to protect people's health while at work.

The National Examination Board in Occupational Safety and Health examinations at both Certificate and Diploma level requires candidates to have a sound understanding of occupational health and hygiene practice and procedures. As with the other 'Guides to Health and Safety Practice', this book has been written particularly with such people in mind.

Whilst aimed at health and safety practitioners, I hope other groups – personnel managers, engineers, lawyers and lecturers, for instance – will find the book helpful.

Jeremy Stranks, 1994

List of abbreviations

ACGIH	American Conference of Government Industrial Hygienists
ACOP	Approved Code of Practice
ACTS	Advisory Committee on Toxic Substances
CHIP	Chemicals (Hazard Information and Packaging for Supply) Regulations 1994
COSHH	Control of Substances Hazardous to Health Regulations 1994
dB	Decibel
DB	Dry bulb temperature
DSE	Display screen equipment
EMA	Employment Medical Adviser
EMAS	Employment Medical Advisory Service
ENA	Employment Nursing Adviser
GT	Globe thermometer temperature
HSC	Health and Safety Commission
HSE	Health and Safety Executive
HSWA	Health and Safety at Work etc. Act 1974
Hz	Hertz
IEE	Institution of Electrical Engineers
ILO	International Labour Organisation
LEV	Local exhaust ventilation
LTEL	Long-term exposure limit
MEL	Maximum exposure limit
mgm^3	Milligrams per cubic metre
ms	Metres per second
mSV	MicroSievert
OEL	Occupational exposure limit
OES	Occupational exposure standard
OHN	Occupational health nurse
PPE	Personal protective equipment
ppm	Parts per million
RCN	Royal College of Nursing
RPE	Respiratory protective equipment
STEL	Short-term exposure limit
TLV	Threshold limit value
WATCH	Working Group on the Assessment of Toxic Chemicals
WB	Natural wet bulb temperature
WBGT	Wet bulb globe temperature
μ	Micron

Principles of occupational health

OCCUPATIONAL HEALTH

'Occupational health' is defined as:

'a branch of preventive medicine concerned with, firstly, the relationship of work to health and, secondly, the effects of work upon the worker'; or

'a branch of medicine concerned with health problems caused by or manifest at work'.

Whilst occupational health practice may incorporate a number of post ill-health activities, such as the treatment of persons suffering ill-health effects as a result of their work or the giving of first aid treatment to people taken ill in the workplace, the principal emphasis is on preventing risks to health of workers arising. Where prevention may not be possible, however, employers must control these risks. This philosophy is inherent in the Control of Substances Hazardous to Health (COSHH) Regulations 1994 and other health-related legislation, such as the Ionising Radiations Regulations 1985 and the Noise at Work Regulations 1989.

Duties of employers

There is a general duty on employers to protect the health of their employees under the Health and Safety at Work etc. Act 1974 (HSWA). Section 2(1) specifies this duty thus:

It shall be the duty of every employer to ensure, so far as is reasonably practicable, the health, *safety and welfare at work of all his employees.*

'Hazard', 'risk' and 'danger'

In any consideration of occupational health strategies it is important to distinguish between the terms 'hazard', 'risk' and 'danger'.

A *hazard* can be defined in a number of ways: 'a situation of risk or danger', 'a situation that may give rise to personal injury', 'the result of a departure from the normal situation which has the potential to cause injury, damage or loss'.

Risk, on the other hand, is defined as 'a chance of loss or injury', 'an exposure to a hazard', 'the probability of a hazard leading to personal injury and the severity of that injury', 'the probability of harm, damage or injury'.

Danger implies 'a thing that causes peril', or 'liability of exposure to harm'.

Health hazards can take many forms, for instance, an incorrectly labelled toxic substance, a badly ventilated mine working or a noisy workshop. It isn't until people come onto the scene, however, that the risks arise.

'Risk' implies an element of uncertainty, the probability that an individual could, for instance, inhale the dust in the mine and contract coal worker's pneumoconiosis or be exposed to sound pressure levels which cause noise-induced hearing loss. Health risks are, therefore, many and varied. Much will depend upon the *dose* of the offending agent, namely the extent or concentration of same related to the duration of exposure, and, in the case of hazardous substances, the *form* taken by that substance, for example a gas, mist, dust or fog. The question of *individual susceptibility* to ill-health conditions must also be considered. No two people necessarily respond in the same way to noise or a hazardous substance for example.

OCCUPATIONAL HEALTH RISKS

Risks to the health of workers can be classified on the basis of physical, chemical and biological risks.

Physical risks

These risks arise as a result of exposure to physical phenomena such as noise, vibration, temperature, various particulate substances, such as dusts, radiation and pressure. Typical examples include: heat – heat cataract, heat stroke; noise – noise-induced hearing loss (occupational deafness); vibration – vibration-induced white finger; dust – silicosis, coal worker's pneumoconiosis; pressure – decompression sickness; job movements – writer's cramp; friction and pressure – beat hand, beat knee, tenosynovitis.

Chemical risks

Exposure to many chemical substances results in various forms of poisoning and other diseases or conditions. These include: acids and alkalis – non-infective dermatitis; metals – lead and mercury poisoning; non-metals – arsenic and phosphorus poisoning; gases – arisine poisoning, carbon monoxide poisoning; organic compounds – occupational cancers, such as bladder cancer.

Biological risks

Ill-health conditions can be caused through exposure to various forms of bacteria, viruses and dusts. Certain diseases are transmissible from animal to man (zoonoses). Biological risks can be classified as: animal-borne – anthrax, brucellosis, glanders; human-borne – viral hepatitis; vegetable-borne – aspergillosis (farmer's lung).

PREVENTION AND/OR CONTROL OF EXPOSURE

Employers have a duty to prevent the exposure of their employees and others to health risks. Where this is not reasonably practicable, they must control that exposure. This duty implies the implementation of a number of strategies which must be related to the degree of risk. Other strategies are available as support strategies. Support strategies, such as the provision and use of personal protective equipment, health surveillance and provision of appropriate welfare amenities, provide extra protection if they are used correctly and on a regular basis.

Prevention strategies

Prevention strategies include:

Prohibition

This is the most extreme form of prevention strategy employed where there is no known form of operator protection available. It implies a total ban on the use of a hazardous substance, process or system where the level of danger is unacceptable.

Elimination

In certain cases it may be possible to eliminate a hazardous substance or working practice completely.

Substitution

The substitution of a less hazardous substance for a more dangerous one is a common prevention strategy. Typical examples include the substitution of toluene for benzene, a notorious carcinogen, and fibreglass for asbestos.

Control strategies

Control strategies include:

Containment (enclosure)

In this case a hazardous substance or environmental stressor, such as noise, is contained or enclosed in order to prevent its liberation into the working environment. Total enclosure is incorporated in many chemical manufacturing processes using a completely sealed system of pipework, processing vessels and tanks. Enclosure may take the form of acoustic enclosure of a noisy process, the use of paint booths and laboratory fume cupboards. In many cases, such enclosures are linked to a local exhaust ventilation system.

Isolation (separation)

Isolation can take a number of forms. For instance, a process using potentially hazardous materials may be located in a part of a premises not frequented by the majority of workers and access to this area may be restricted. Similarly, a nuclear reactor could be located in a remote part of the countryside well away from centres of population.

Ventilation

Infiltration of air into buildings through openings in the fabric and even **planned natural ventilation** give no continuing protection wherever toxic fumes, gases, vapours, etc. may be present. **Local exhaust ventilation** (LEV) systems must, in most cases, be operated. In certain limited cases, **dilution ventilation** systems may be satisfactory. (See further Chapter 7.)

Segregation

Segregation is a method of controlling the risks from hazardous substances or physical hazards such as noise and radiation. It can take a number of forms, as follows.

Segregation by distance (separation)

This is a process whereby an individual separates himself from the source of the danger. This applies in the case of noise where, as the distance from the source increases, so the risk of occupational deafness reduces. Similar principles apply to radiation. Segregation by distance protects those at secondary risk, if those at primary risk are protected by other forms of control.

Segregation by age

The need for the protection of young workers has reduced significantly but, where the risk is minimal it may be necessary to exclude young persons, particularly females, from an activity. An example of such segregation is to be found in the Control of Lead at Work Regulations 1980, which exclude the employment of young persons in lead processes.

Segregation by time

This refers to the restriction of certain hazardous operations to periods when the number of workers present is small, for instance at night or during weekends, and when the only workers at risk are those involved in the operation. An example of segregation by time is the examination by radiation of very large castings.

Segregation by sex

There is always the possibility of sex-linked vulnerability to certain toxic materials, particularly in the case of pregnant women, where there can be damage to the foetus e.g. in certain processes involving lead.

Change of process

Improved design or process engineering can bring about changes to provide better operator protection. This is appropriate in the case of dusty processes or machinery noise.

Controlled operation

Controlled operation is closely allied to the duty under HSWA section 2(2)(c), to provide a safe system of work. It is particularly appropriate where there is a high degree of foreseeable risk. It implies the need for high standards of training, instruction, supervision and control, and may take the following forms:

- isolation of processes in which hazardous substances are used or where there may be a risk of, for instance, heat stroke
- the use of mechanical or remote control handling systems e.g. with radioactive substances
- the use of permit to work systems e.g. entry into confined spaces, such as closed vessels, tanks and silos, or fumigation processes using hazardous substances such as methyl bromide
- restriction of certain activities to highly trained, skilled and supervised staff e.g. competent persons working in high-voltage switchrooms.

Reduced time exposure (limitation)

Risks to health from hazardous substances or physical stressors such as noise, can be reduced by limiting the exposure of workers to certain predetermined maxima. The strategy forms the basis for the establishment of occupational exposure limits i.e. long-term exposure limits (8-hour time-weighted average value) and short-term exposure limits (10-minute time-weighted average value). This strategy is encompassed in the Noise at Work Regulations 1989.

Dilution

Danger may arise from the handling and transport of hazardous substances in concentrated form. Diluting a concentrated substance with water, and thereby reducing the risks, is commonly practised in the transportation of certain hazardous wastes, or where substances may be fed into processing plant. Dilution as a control strategy has very limited application, however.

Neutralisation

This is the process of adding a neutralising compound to another strong chemical compound e.g. acid to alkali or vice versa, thereby reducing the immediate danger. As with dilution, this strategy is commonly used in the transportation of hazardous substances and wastes, such as acid-based wastes, and in the treatment of processing plant effluents prior to their discharge to a public sewer.

Support strategies

Support strategies are important in that they assist the various prevention and control strategies. Support strategies include:

Cleaning and housekeeping arrangements

Poor levels of cleaning and housekeeping are a contributory factor in the causes of accidents and occupational ill health. The cleaning function should be managed through the use of a formal cleaning schedule or programme. Such a schedule should identify each area, item of plant, structural finish, etc. which requires cleaning and the method, materials and equipment to be used for the cleaning operation. The schedule should also incorporate, in each case, the frequency of cleaning, the individual responsible for ensuring the cleaning task is completed satisfactorily and any precautions that may be necessary, such as the use of certain items of personal protective equipment when mixing cleaning chemicals prior to use. Emphasis should be placed on the use of mechanical cleaning equipment, rather than brushes, brooms, string mops, etc.

Housekeeping inspections should be undertaken regularly, particularly in chemical storage and handling areas.

Preventive maintenance

The absence of a planned maintenance procedure for plant and equipment can result in deterioration of plant efficiency and an increased risk to operators of contact with hazardous substances. A formal planned maintenance procedure should identify each item of plant, the maintenance procedure to be followed, the frequency of such maintenance, individual responsibility for ensuring the maintenance tasks are completed satisfactorily, and any precautions necessary on the part of maintenance personnel, such as the use of eye protection, plant isolation procedures, etc.

Welfare amenity provisions

There is a legal duty on employers to provide appropriate welfare amenity provisions, namely sanitation, hand washing, shower, drinking water, clothing storage and eating facilities for staff under the Workplace (Health, Safety and Welfare) Regulations 1992. Such facilities, and the correct use of same by staff, are an important aid in the prevention of personal contamination from hazardous substances and the risk of occupational diseases that could result from such contamination e.g. dermatitis.

Personal hygiene

A high level of personal hygiene is required by operators involved in the use of hazardous substances. Personal hygiene control measures include:

- strict control over decontamination procedures, particularly before eating, drinking, smoking or leaving the workplace at termination of work
- a prohibition on eating, drinking and smoking wherever there may be risk

of hand-to-mouth contamination e.g. handling hazardous substances

- the use of barrier creams and other forms of skin protection directly related to the risks
- procedures for sanitisation of face, eye and hearing protection, or the use of disposable forms of protection
- a total ban on the practice of workers returning home while wearing contaminated protective clothing and footwear
- training at induction and on a regular basis in the principles of personal hygiene and its relationship to the health risks
- linking personal hygiene requirements with current industrial relations policy
- the formulation and formalisation of a company hygiene policy accompanied by strict enforcement.

Much will depend upon the level of risk associated with poor personal hygiene as to whether all or some of the above measures are necessary.

Personal protective equipment

The provision and use of any form of personal protective equipment, such as eye protection, hand protection, respiratory protection, is an important support strategy in the prevention and control of health risks. However, it is not the perfect solution to prevention and control and should be viewed mainly as an extra form of protection. In the majority of cases, it should be viewed as the last resort, when all other measures have failed, in protecting people at work.

Moreover, reliance on personal protective equipment as a sole means of protecting workers from health risks could have tragic consequences due to the reluctance of workers to wear the items of personal protective equipment for 100 per cent of the time that they may be exposed to these risks. (See further Chapter 8.)

Health surveillance

Health surveillance implies the regular monitoring of the state of health of individuals through the use of a number of techniques aimed at detecting the presence of hazardous substances in the body and other evidence of the early stages of occupational ill health. Health surveillance concentrates particularly on two groups of workers:

1. those at risk of developing further ill health or disability by virtue of the present state of health; and
2. those actually or potentially at risk by virtue of the type of work they undertake during their employment e.g. lead workers.

(See further Chapter 6.)

First aid

The Health and Safety (First Aid) Regulations 1981 and Approved Code of Practice (ACOP) lay down the standards for the provision of first aid treatment. First aid procedures should include those for dealing with potential contamination situations.

Information, instruction and training

Staff should be informed of the health risks and instructed in the precautions necessary when exposed to such risks. Training in the procedures to prevent or control health risks should be provided at induction, on change of process and/or substances and on a regular refresher basis.

Propaganda

The use of various forms of propaganda, aimed at drawing the attention of operators to identified health risks, is a continuing process. A wide range of posters, videos, films and other forms of propaganda is available.

Joint consultation

There is a legal duty on the part of employers to consult with employees, or their representatives, on health risks and the precautions necessary. Joint consultation may take place between management and trade union appointed safety representatives and/or through the operation of a health and safety committee.

OCCUPATIONAL HEALTH PRACTITIONERS

Practitioners in occupational health include occupational health nurses, occupational physicians, occupational hygienists, ergonomists and health and safety specialists. Each of these groups has an important role in protecting the health of workers.

Occupational health nurses

An occupational health nurse (OHN) should preferably hold a recognised qualification in occupational health nursing, such as the Occupational Health Nursing Certificate of the Royal College of Nursing (RCN), in addition to a general nursing qualification i.e. RGN.

The OHN's role consists of eight main elements:

- health supervision
- health education
- environmental monitoring and occupational safety
- counselling
- treatment services

- rehabilitation and resettlement
- unit administration and record systems
- liaison with other agencies.

These functions are capable of achievement in the following ways:

1 *Acting for or doing for another* This function is concerned with assisting people with tasks that they cannot do for themselves, such as helping disabled workers with certain tasks or performing certain forms of treatment.
2 *Guiding* Health education is a continuous process. Guidance must also be given to workers returning to work following illness or injury, and to special groups such as young persons, pregnant women and disabled workers.
3 *Supporting physically and psychologically* Counselling of workers on a wide range of subjects, many of which are not necessarily health-related, is an important feature of the OHN's role.
4 *Providing the right environment* Acting independently or as a member of a health and safety team, the OHN has an important role to play in examining the overall work environment, particularly where there may be risks to health or early signs of health deterioration among workers.
5 *Teaching* Teaching is largely aimed at the prevention of accidents and occupational ill health. The OHN, through her one-to-one relationship with workers, can teach them to avoid hazards. In many organisations, the training of first aid staff is an important part of this teaching function.

The RCN identifies the following duties which a fully trained OHN could perform:

- health assessment in relation to the individual worker and the job to be performed
- noting normal standards of health and fitness and any departures or variations from these standards
- referring to the occupational physician or doctor such cases which, in the opinion of the nurse, require further investigation and medical, as distinct from nursing, assessment
- health supervision of vulnerable groups e.g. young persons, disabled workers
- routine visits to and surveys of the working environment, and informing as necessary the appropriate expert when a particular problem requires further specialised investigation
- employee health counselling
- health education activities in relation to groups of workers
- the assessment of injuries or illness occurring at work and treatment or referral as appropriate

- responsibility for the organisation and administration of occupational health services, and the control and safe-keeping of non-statutory personal health records
- a teaching role in respect of the training of first aid personnel and the organisation of emergency services.

The RCN further recommends that where a full-time medical officer with occupational health training and management is employed, the doctor assumes overall responsibility for the leadership and organisation of the occupational health service. As a matter of principle in such organisations, nursing staff organised in hierarchies work to one nurse leader who is responsible for the overall organisation and administration of the occupational health nursing services. The most senior nurse should work in close partnership with the doctor in charge.

Occupational physicians

Occupational medicine is a branch of preventive medicine concerned mainly with the diagnosis and treatment of occupational diseases and conditions. An occupational physician should preferably hold a recognised qualification in occupational medicine in addition to being a registered medical practitioner. The British Medical Association (BMA) has identified the duties and responsibilities of doctors holding appointments in occupational medicine as encompassing the following:

1 **The effects of health on the capacity to work**
This aspect includes:

- the provision of advice to employees on all health matters relating to their working capacity
- examination of applicants for employment and advice as to their placement
- immediate treatment of medical and surgical emergencies occurring at the place of employment
- examination and continued observation of persons returning to work after absence due to illness or accident and advice on suitable work
- health supervision of all employees with special reference to young persons, pregnant women, elderly persons and disabled persons.

2 **The effects of work on health**
This aspect of the occupational physician's role includes:

- responsibility for nursing and first aid services
- the study of the work and working environment and their effects on the health of employees
- periodical examination of persons exposed to special hazards in respect of their employment

- advice to management regarding:
  — the working environment in relation to health
  — occurrence and significance of hazards
  — accident prevention
  — statutory requirements in relation to health
- medical supervision of the health and hygiene of staff and facilities with particular reference to canteens, kitchens, etc. and those working in the production of foods or drugs for sale to the public
- the arranging and carrying out of such education work in respect of the health, fitness and hygiene of employees as may be desirable and practicable
- advising those committees within the organisation which are responsible for the health, safety and welfare of employees.

Occupational hygienists

Occupational hygiene is concerned with the identification, measurement, evaluation and control of contaminants and physical phenomena, such as noise and radiation, which would otherwise have unacceptable adverse effects on the health of people exposed to them.

The occupational hygienist, therefore, is fundamentally involved with the practical measurement, monitoring, evaluation and control of health risks to employees. These risks can include those associated with noise, vibration, radiation, toxic substances and other forms of environmental or occupational stress. Control methods specified may be based on analysis of the results of environmental monitoring operations and current hygiene standards e.g. occupational exposure limits. (See further Chapter 6.)

Entry to the profession is controlled by the British Examination and Registration Board in Occupational Hygiene (BERBOH).

Health and safety practitioners

The role, function, level of training and responsibilities of health and safety practitioners vary enormously. Broadly, anyone can be appointed as an organisation's health and safety specialist. Health and safety practitioners include full-time health and safety managers, health and safety advisers, safety officers and safety engineers. At the other end of the scale are a wide range of people who are responsible for the health and safety function along with their principal role as, for instance, personnel manager, company secretary or company engineer.

However, the Management of Health and Safety at Work Regulations 1992 require an employer 'to make a suitable and sufficient assessment of the risks to the health and safety of his employees and other persons not in his employment' (regulation 3). On the basis of this risk assessment, he must 'appoint one or more competent persons to assist him in undertaking the measures he needs to take to comply with the requirements and prohibitions imposed upon him by or under the relevant statutory provisions' (regulation 6).

Health and safety practitioners have an important role in occupational health practice. This may include the development of measures to comply with the Control of Asbestos at Work Regulations 1987, COSHH Regulations 1994 or Ionising Radiations Regulations 1985. In some cases, they operate on a team basis with OHNs, occupational physicians and occupational hygienists. In other cases they may undertake the hygienist function personally, liaising with external occupational health practitioners.

OCCUPATIONAL HEALTH AUTHORITIES

The principal occupational health authority is the Employment Medical Advisory Service (EMAS). This organisation was established under the Employment Medical Advisory Service Act 1972 and the principal part of the service is the Medical Services Division of the HSE. This consists of a national network of doctors (Employment Medical Advisers) and nurses (Employment Nursing Advisers) accountable to Senior Employment Medical Advisers. in addition, many examinations are undertaken by doctors who have been appointed for this purpose by EMAS (appointed doctors).

Employment medical advisers (EMAs) carry out statutory examinations of young persons and other persons, some of which are of a compulsory nature under specific legislation, such as the Control of Lead at Work Regulations 1980 and the Diving Operations at Work Regulations 1981.

EMAs have a number of statutory powers. For instance, section 10 of the Factories Act 1961 gives an EMA power to order the medical examination of factory workers. Where, in the opinion of an EMA, the health of an employee has been, is being or will be injured, due to the nature of his work, he may serve a written notice on the employer requiring him to permit a medical examination of that employee.

In certain hazardous occupations, periodic medical examination and/or supervision of employees is compulsory. This includes, for example:

- persons employed in work involving compressed air operations (Work in Compressed Air Special Regulations 1958)
- persons employed in lead processes (Control of Lead at Work Regulations 1980)
- persons employed in processes involving ionising radiation (Ionising Radiation Regulations 1985)
- persons employed in underwater diving operations (Diving Operations at Work Regulations 1981).

Moreover, a health risk assessment made under the COSHH Regulations 1994 may require some form of medical surveillance.

PRE-EMPLOYMENT HEALTH SCREENING QUESTIONNAIRE

Surname ______________________ **Location** ______________________

Forenames ______________________ **Date of birth** ______________________

Address ____

____

Tel. no. ______________________ **Occupation** ______________________

Position applied for ____

Name and address of doctor ____

____

SECTION A

Please tick if you are at present suffering from, or have suffered from:

1	Giddiness	☐	8	Stroke	☐
	Fainting attacks	☐		Heart trouble	☐
	Epilepsy	☐		High blood pressure	☐
	Fits or blackouts	☐		Varicose veins	☐
2	Mental illness	☐	9	Diabetes	☐
	Anxiety or depression	☐	10	Skin trouble	☐
3	Recurring headaches	☐	11	Ear trouble or deafness	☐
4	Serious injury	☐	12	Eye trouble	☐
	Serious operations	☐		Defective vision (not	☐
5	Severe hay fever	☐		corrected by glasses or	☐
	Asthma	☐		contact lenses)	☐
	Recurring chest disease	☐		Defective colour vision	☐
6	Recurring stomach trouble	☐	13	Back trouble	☐
	recurring bowel trouble	☐		Muscle or joint trouble	☐
7	Recurring bladder trouble	☐	14	Hernia/rupture	☐

SECTION B

Please tick if you have any disabilities that affect:

☐ Standing	☐ Lifting	☐ Working at heights
☐ Walking	☐ Use of your hands	☐ Climbing ladders
☐ Stair climbing	☐ Driving a vehicle	☐ Working on staging

SECTION C

How many working days have you lost during the last three years due to illness or injury? ____________days

Are you at present having any tablets, medicine or injections prescribed by a doctor? YES/NO

Are you a registered disabled person? YES/NO

• **FIG 1.1 Health questionnaire**

SECTION D

Previous occupations	Duration	Name & address of employer

SECTION E

The answers to the above questions are accurate to the best of my knowledge.

I acknowledge that failure to disclose information may require reassessment of my fitness and could lead to termination of employment.

Signature ________________ Prospective employee Date______________

Signature ________________ Manager Date______________

FIG 1.1 continued

PRINCIPAL AREAS OF OCCUPATIONAL HEALTH PRACTICE

Placing people in suitable work

The aim is to assess current mental and physical capability and to identify pre-existing ill-health conditions. This generally takes the form of:

- pre-employment medical examinations; and/or
- pre-employment health screening. (See Fig 1.1)

Health surveillance

This entails specific health examinations at a predetermined frequency for:

- those at risk of developing further ill health or disability e.g. people exposed to excessive noise levels; and
- those actually or potentially at risk by virtue of the type of work they undertake during their employment e.g. radiation workers.

Providing a treatment service

The efficient and speedy treatment of injuries, acute poisonings and minor ailments at work is a standard feature of occupational health practice. This service is important in terms of keeping people at work, thereby reducing lost time associated with attendance at local doctors' surgeries and hospital casualty departments.

Primary and secondary monitoring

Primary monitoring is largely concerned with the clinical observation of sick people who may seek treatment or advice on their condition. Secondary monitoring is directed at controlling the hazards to health which have already been recognised e.g. audiometry.

Avoiding potential risks

This is an important feature of occupational health practice with the principal emphasis on prevention, in preference to treatment, for a known condition.

Supervision of vulnerable groups

Vulnerable workers include young persons, pregnant women, the aged, the disabled and people who may have long periods of sickness absence. Routine health examinations to assess continuing fitness for work may be incorporated.

Monitoring for early evidence of non-occupational disease

This entails the routine monitoring of workers not exposed to health risks with the principal objective of controlling diseases prevalent in certain communities e.g. mining, with a view to their eventual eradication.

Counselling

Counselling of people on health-related issues and on personal, social and emotional problems is a standard area of occupational health practice.

Health education

This is primarily concerned with the education of employees towards a healthier lifestyle. It can also include training of management and staff in their respective responsibilities for health and safety at work, in healthy working techniques and in the avoidance of health risks.

First aid and emergency services

This area may include the supervision of first aid arrangements, training of first aid staff and the preparation of contingency arrangements in the event of fire, explosion or other potential emergency situations.

Environmental control and occupational hygiene

This entails the identification, measurement, evaluation and control of potential health risks to employees and the general public from processes and premises, together with the prevention and control of environmental nuisance e.g. from dust, noise, etc.

Liaison

Occupational health staff maintain liaison with enforcement agency staff, such as employment medical advisers and employment nursing advisers, and other public agencies.

Health records

The completion and maintenance of health records is required under certain regulations e.g. the Noise at Work Regulations 1989, Control of Substances Hazardous to Health Regulations, 1994, and for internal purposes to ensure continuing surveillance procedures are maintained. Health records may feature in epidemiological studies of groups of workers.

Chemical health hazards

CLASSIFICATION OF HAZARDOUS CHEMICAL SUBSTANCES AND PREPARATIONS

The **Chemicals (Hazard Information and Packaging for Supply) Regulations 1994** are the principal legislation concerned with the classification of hazardous chemical substances and preparations. The classification of such substances and preparations hazardous for supply is based on their physico-chemical properties and effects on health, together with the effects on the environment.

The purpose of classification is to identify the properties of substances and preparations that may constitute a hazard during normal handling and use.

Classified substances must display a supply label which is intended to provide a primary means by which people at work and the general public are given essential information about hazardous substances and preparations (see Fig 2.1). The label draws the attention of users to the inherent hazards of such materials so that the necessary precautionary measures can be taken. The label may also serve to draw attention to more comprehensive product information on safety such as the supplier's safety data sheet.

The label should take account of all potential hazards that are likely to arise in normal handling and the use of a dangerous substance or preparation in the form in which it is supplied, although not necessarily in any different form in which it may ultimately be used, e.g. diluted. The information on the label should include, in relation to both substances and preparations dangerous for supply (regulation 9 of CHIP Regulations):

- the name, full address and telephone number of a person in a member state who is responsible for supplying the substance or preparation
- the indication or indications of danger and the corresponding symbol or symbols
- the risk phrases (set out in full)
- the safety phrases (set out in full).

Further specific information is required depending on whether the chemical is a substance or preparation. Details of this classification and labelling procedure are contained in the Approved Guide to the Classification and Labelling of Substances and Preparations Dangerous for Supply (the Approved Guide) issued with the Regulations.

In the Approved Guide the criteria for classification, choice of symbols, indication of danger and choice of risk phrases are grouped under the following categories:

- **Physico-chemical properties** i.e. explosive, oxidising, extremely flammable, highly flammable and flammable
- **Health effects** i.e. very toxic, toxic, harmful, corrosive, irritant, sensitising, carcinogenic, mutagenic, toxic for reproduction and dangerous for the environment.

The requirements of the CHIP Regulations are dealt with in Chapter 9.

STORAGE AND HANDLING OF HAZARDOUS CHEMICALS

Prior to storing and handling hazardous chemicals, it is imperative to consult sources of safety data from the manufacturers and suppliers of the chemical substances and preparations. (Safety data should comply with the Approved Code of Practice on Safety Data Sheets Dangerous for Supply issued with the CHIP Regulations.)

General precautions

The following precautions are necessary to ensure the safe handling and storage of dangerous chemical substances and preparations:

- meticulous standards of housekeeping should be maintained at all times
- smoking and the consumption of food or drink should be prohibited in any area where substances are used or stored e.g. laboratory, bulk chemical store
- staff must be reminded regularly of the need for good personal hygiene, in particular washing of hands after handling chemical substances
- minimum quantities only should be stored in the working area; extra bulk storage may be required well away from the work area
- containers and transfer containers should be clearly and accurately marked
- chemical substances should always be handled with care and carriers used for Winchester and other large containers
- fume cupboards should operate with a minimum face velocity of approximately 0.4 m/sec when measured with the sash opening set at 300mm maximum, and performance should be checked frequently in accordance with the COSHH Regulations

- staff should always wear personal protective clothing and equipment e.g. eye protection, face protection, aprons, gloves, wellington boots, whenever handling or using dangerous chemical substances
- any injury should be treated promptly, particularly skin wounds and abrasions
- responsibility for safe working should be identified at senior management level, and written procedures published and used in the training of staff.

Precautions with specific substances

Flammable liquids

- All containers should be of the self-closing type. Caps should be replaced after dispensing. Liquids should be dispensed over a drip tray.
- Containers should be stored in a well-ventilated fire-protected area.
- Fire appliances should be located in a readily accessible position and staff trained in their use.
- Flammable liquids should be transported in closed containers of metal construction. (Some plastics may, however, be acceptable for this purpose.)

Carcinogens, poisons, etc.

- Staff must wear the appropriate protective clothing and equipment.
- First aid treatment, including the appropriate antidote, must be known and readily available.
- Substances producing fumes must be handled in fume cupboards.
- Substances should be transported in sealed and labelled containers

Solids

In the case of dusts and other particulate material, such as powders:

- respiratory protection should be worn unless control measures are adequate
- atmospheric concentrations should be measured and related to the current hygiene standard in order to determine the degree of danger present
- there should be a complete ban on smoking and naked lights where the solid is flammable.

In the case of all solids, the nature of the substance to be handled must be ascertained.

Bulk storage of hazardous chemical substances

In the design and use of bulk storage facilities, the following aspects need attention:

IDENTIFYING CHEMICALS

- CHECK LABELS
- CHECK SYMBOLS
- CHECK CONTAINERS

} DO THEY TIE UP?

HAZARD WARNING SYMBOLS

CORROSIVE

OXIDISING

EXPLOSIVE

TOXIC

HARMFUL/ IRRITANT

HIGHLY FLAMMABLE

IN CASE OF DOUBT –

DO NOT TOUCH – ASK!

• **FIG 2.1 Chemical hazard warning symbols**

(a) the range and quantities of chemicals to be stored
(b) dependent upon (a) above, the degree of segregation by distance of:
  — the store from any other building; and
  — certain chemical substances within the store from other chemical substances stored.

Segregation

The aim in segregating stored chemicals should be:

- to facilitate emergency access and escape in the event of fire or other emergency
- to separate incompatible chemicals to prevent their mixture e.g. by spillage, or wetting during cleaning processes
- to separate process areas, which normally contain relatively small quantities, from storage areas containing larger amounts
- to prevent rapid fire spread, or the evolution of smoke and gases which can be produced in a fire
- to isolate oxidising agents which, when heated, will enhance a fire, perhaps to explosive condition
- to isolate those substances which decompose explosively when heated
- to minimise toxic hazards arising from loss of containment through spillage, seepage or package deterioration
- to minimise the risk of physical damage e.g. by fork-lift trucks, to containers
- to separate materials where the appropriate fire-fighting medium for one may be ineffective for, or cause adverse reaction with, another.

Structural requirements

Chemical storage may take two forms, namely an open area or a purpose-built store. Open storage is not recommended but, when it is unavoidable, it should comprise a secure area fenced to a height of 2m with a lockable access point.

Purpose-built chemical stores should be of the detached single-storey brick-built type or constructed in other suitable materials, such as concrete panels, with a sloping roof of weatherproof construction. The structure should have a notional period of fire resistance of at least one hour.

Other features should include:

- permanent ventilation by high- and low-level air bricks set in all elevations, except in those forming a boundary wall; low-level air bricks should be sited above door sill level
- access doors constructed from material with at least one hour notional period of fire resistance; doorways should be large enough to provide access for fork-lift trucks, with ramps on each side of the door sill; separate pedestrian access, which also serves as a secondary means of escape, should be provided

- an impervious chemical-resistant finish to walls, floors and other surfaces
- artificial lighting by sealed bulkhead or fluorescent fittings, to provide an overall illuminance level of 300 lux.

General requirements

In the case of both open and closed storage, the following are required:

- provision of adequate space, with physical separation and containment for incompatible substances, each area to be marked with the permitted contents, the hazards and the necessary precautions, and incorporating an area for the storage of empty containers
- fire separation of individual areas sufficient to prevent fire spreading
- provision of the following equipment in a protected area outside the store:
  - fire appliances – dry powder and/or foam extinguishers
  - fixed hose reel appliance
  - emergency shower and eyewash station with water heating facility to prevent freezing
  - personal protective equipment i.e. safety helmet with visor, impervious gloves, disposable chemical-resistant overall, with storage facilities for same
  - respirator and breathing apparatus in a marked enclosure
- a total prohibition on the use of naked flames and smoking; appropriate warning signs should be displayed
- a prohibition on the use of the storage of other items
- provision of racking or pallets to enable goods to be stored clear of the floor.

Storage system

The system for storage must be simple and compatible with legal requirements for classification and labelling.

TRANSPORT OF HAZARDOUS CHEMICALS

Employers have a general duty under HSWA to protect members of the public from hazards arising from their activities. This duty applies particularly in the case of the transport of hazardous substances by road and is reinforced by the Dangerous Substances (Conveyance by Road in Road Tankers and Tank Containers) Regulations 1981. These regulations apply to the conveyance of dangerous chemicals by road in a tanker or tank container, including any loading and unloading activities at premises. The regulations incorporate an **Approved List** of dangerous substances (approved substance identification numbers, emergency action codes and classifications for dangerous substances conveyed in road tankers and tank containers).

A **dangerous substance** is defined with reference to the Approved List as:

'any substance (including any preparation) which is either contained in Part 1 of the Approved List (unless it is in such diluted form as not to create a risk) or any other substance which by reason of its characteristic properties creates a risk to the health and safety of any person in the course of conveyance by road, which is comparable with the risk created by substances which are specified in the Approved List'.

The characteristic properties are listed in Schedule 1. An ACOP, 'Classification of Dangerous Substances for Conveyance in Road Tankers and Tank Containers', provides practical guidance on how substances not on the Approved List can be classified to ascertain whether or not they come within the regulations.

Vehicles and tanks must be properly designed, of adequate strength and of good construction from sound and suitable materials before they can be used to convey dangerous substances by road. Operators must bear in mind the nature and circumstances of the journey and the characteristic properties and the quantity of substances being carried. Provision is made for the testing and examination of tanks and tank containers.

Specific duties are placed on operators and drivers. Operators must ensure they are aware of the risks by obtaining from the consignor or others relevant information on the dangerous substances. They must ensure that drivers are informed in writing of the identity of the substance and the nature of the dangers to which the substance could give rise, together with the emergency action which should be taken in appropriate circumstances. Operators must provide adequate instruction and training for drivers, together with documentation of this training.

Drivers must ensure that vehicles, when not in use, are safely parked or supervised when they are carrying prescribed substances. They must also take the precautions necessary for the prevention of fire or explosion.

Tanker and tank container marking

All road tankers and tank containers must carry at least two hazard warning panels to the specification outlined in Schedule 4 of the regulations. Panels must be weather-resistant, indelibly marked and rigidly fixed. Details of the form and specification of hazard warning panels are outlined in Fig 2.2.

The form of the hazard warning panel is indicated in (a) and (b) of Fig 2.2. The allocation of the five spaces, as numbered in Fig 2.2 (a), is as follows:

(1) Emergency action code.
(2) Substance identification number and, if included, the name or, in the case of multi-loads, the word 'multi-load'.
(3) Hazard warning symbol.
(4) Telephone number or other approved text.
(5) Optional manufacturer's or owner's name or house symbol or both.

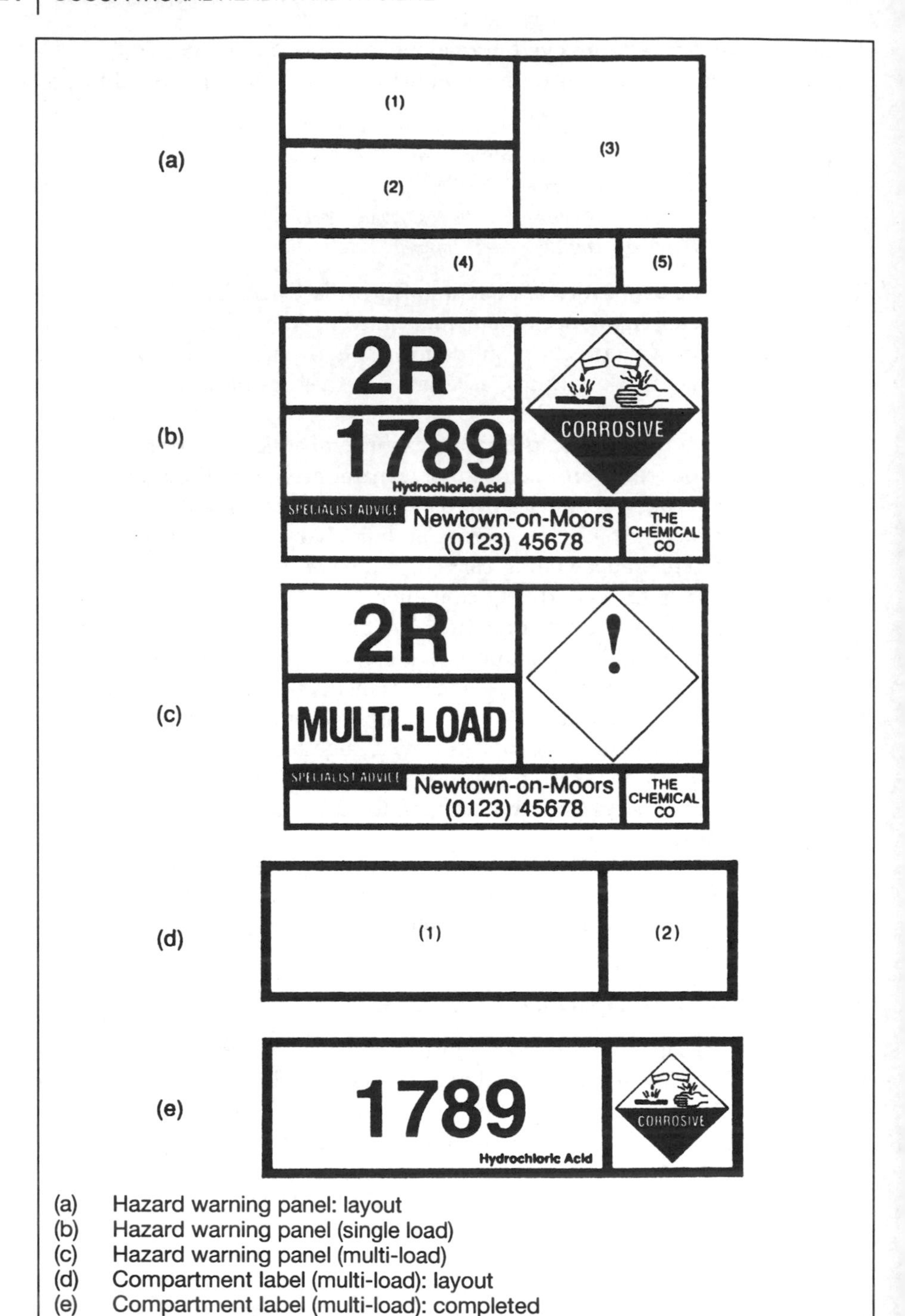

(a) Hazard warning panel: layout
(b) Hazard warning panel (single load)
(c) Hazard warning panel (multi-load)
(d) Compartment label (multi-load): layout
(e) Compartment label (multi-load): completed

• **FIG 2.2 Hazard warning panels and compartment labels** as required by the Dangerous Substances (Conveyance by Road in Road Tankers and Tank Containers) Regulations 1981

The colour of the hazard warning panel must be **orange** and conform to the specification for that colour in Part 2 of Schedule 4 of the regulations, except that the space for the hazard warning sign must be **white**, and the borders, internal dividing lines, letters and figures, **black**.

Where the emergency action code or the multi-load emergency action code, ascertained from Schedule 1 or 2, is a **white** number and/or letter on a **black** background, it/they must be displayed on the panel as **orange** on a **black** background. The letters must appear in a **black** rectangle having the height of 100mm and a width of 10mm greater than the width of the letter.

The form of the compartment label for multi-loads of substances with different hazards is set out in Fig 2.2, and the spaces must be used for the following purposes:

(1) substance identification number and, if included, name

(2) hazard warning symbol.

An example is shown at Fig 2.2(e).

Where the multi-load consists of substances subject to the same hazards, the square labelled (2) in Fig 2.2(d) can be omitted from the compartment label.

The colour of the compartment label must be **orange** and conform to the specification for that colour in Part 2 of Schedule 4, except for the space for the hazard warning symbol (where one is required), which must be **white** and the borders **black**.

DISPOSAL OF HAZARDOUS SUBSTANCES

The Environmental Protection Act (EPA) 1990 is the principal legislation dealing with the disposal of both domestic and industrial waste. Procedures for the safe disposal of dangerous substances depend largely on the type and quantity of material involved. In all cases the local waste disposal authority must be consulted since disposal facilities vary considerably in different areas of the country.

The following must be considered prior to disposal of dangerous substances or their by-products:

- Only trained and authorised staff should be permitted to dispose of dangerous substances. Where large quantities are concerned, disposal should be carried out in conjunction with the waste disposal authority.
- Where contractors provide a waste disposal service, it is necessary to know:
  - the location of the disposal site
  - the procedure for disposal
  - whether formal licensing is required for the disposal
  - the mode of transport of the dangerous substances from the premises to the disposal site.

CONTROL STRATEGIES FOR CHEMICAL HEALTH HAZARDS

The effect on the body of exposure to hazardous chemical substances vary enormously. These effects, and the control strategies necessary, for a number of well-known substances are dealt with below.

Lead

Lead is, perhaps, the oldest known metal, its use being common in the Roman age and thereafter. Poisoning by lead or a compound of lead as a result of the use or handling of, or exposure to the fumes, dust or vapour of, lead or a compound of lead, or a substance containing lead, is a prescribed occupational disease.

Lead is mined principally as lead sulphide (galena) in many countries and is used extensively throughout the world as a component of electric batteries, cables, as lead sheet and pipe, as an anti-detonant in petrol, in solders and alloys and as a constituent of paints, glass and pottery. As such, the use and working of lead features in many industries.

Legal requirements relating to the protection of lead workers are covered by the Control of Lead at Work Regulations (CLWR) 1980 and accompanying ACOP. (See further Chapter 9.)

Effects of exposure

Lead may enter the body by inhalation of the fumes or dust, by ingestion of contaminated food or through pervasion of the unbroken skin in the form of organic compounds such as tetraethyl lead. Lead is a cumulative poison. In the case of chronic absorption it is deposited in the calcareous portion of the bones as an insoluble and harmless triple phosphate.

Poisoning takes two distinct forms. The common form arises from handling metallic lead and its inorganic compounds and the second (rarer) form from handling alkyl organic compounds. In acute cases of inorganic lead poisoning, usually caused by exposure to fumes, the initial symptoms are a sweetish taste in the mouth, especially on smoking, with anorexia, nausea, vomiting and headache, sometimes persistent constipation and intermittent colic. In more chronic poisoning, the classical symptoms and signs are headache, pallor, a blue line at the gum margins, anaemia and palsy (wrist and foot drop), together with certain biochemical abnormalities and encephalopathy, the latter being characterised by mental dullness, inability to concentrate, faulty memory, tremors, deafness, convulsions and coma.

In the majority of organic lead poisonings, exposure to tetraethyl lead is the cause. This particular compound is volatile at room temperature and is readily inhaled. On absorption into the body it exercises a focal effect on the central nervous system. In severe cases of poisoning the effect on the central nervous system is startling, commencing with general malaise but progressing to talkativeness, excitement and muscular twitchings, with insomnia, delusions, hallucinations, and even acute and violent mania.

Control strategies

Controls are directed at preventing exposure to lead, its compounds, alloys or as a constituent of any substance, and by means other than by the use by operators of respiratory protective equipment and protective clothing. An employer must assess the risk of exposure before work is commenced with a view to determining the nature and potential degree of exposure.

Measures to prevent or control exposure include:

- *Substitution* the use of lead-free materials or low-solubility lead compounds.
- *Change in form* the use of lead or lead compounds in the form of an emulsion or paste.
- *Temperature control* the operation of processes at below 500°C to reduce or eliminate fume emission.
- *Containment* operation of a process in enclosed plant.
- *LEV systems* where containment is not reasonably practicable, the use of effective LEV systems.
- *Wet methods* use of wet grinding techniques and wetting of floors and workbenches.

Such measures should be accompanied by information, instruction and training for operators and rigid controls over eating, drinking and smoking in the workplace, separation of protective clothing from personal clothing, a high standard of welfare amenity provision and personal protective equipment. These control measures must be supported by meticulous levels of cleaning and housekeeping.

Air monitoring is also required on a regular basis, together with medical surveillance of potentially exposed employees.

Benzene

Benzene is a freely volatile solvent with a distinctive odour and narcotic action. It is one of the products of the destructive distillation of coal tar. Crude benzene, such as substances known as 'benzol' or 'benzole', frequently contains as impurities varying amounts of the homologues of benzene, for instance, toluene and xylene. It may also contain carbon bisulphide, phenol and other coal tar derivatives. Included in the benzene family are benzene (benzole), nitrobenzene and aniline.

Poisoning by benzene or a homologue of benzene through the use or handling of, or exposure to the fumes of, or vapour containing benzene or any of its homologues, is a prescribed occupational disease.

Effects of exposure

Poisoning may be acute or chronic. In its initial stages acute benzene poisoning gives rise to a feeling of euphoria, or merely to headache, giddiness, nausea and vomiting. In the more serious cases, excitement with a sensation of constriction in the chest, convulsive movements and paralysis may give way, as intoxication increases, to delirium, coma and death due to respiratory failure. Such cases generally arise as a result of accidental exposure to high concentrations of fumes, perhaps following a spillage. Toluene and xylene have a similar but less toxic effect.

Chronic benzene poisoning is more common and affects the bone marrow causing anaemia. Early symptoms include lassitude, muscular weakness, mild digestive disturbances, pallor and giddiness. Subsequent symptoms include severe anaemia, with extreme pallor, muscular weaknesses, haemorrhages from the mucous membranes and skin haemorrhages. Death my result from aplastic anaemia.

Control strategies

Benzene has virtually been banned from use in industry, other than for use in specific research operations. The principal control strategy is that of substitution of benzene by the safer toluene.

Trichloroethylene

Trichloroethylene, commonly known as 'Trike', 'Triklone' and 'Trilene', has widespread use, either in the pure state as an anaesthetic or mixed with other solvents. Its main uses are as a degreasing agent for metals, as a dry cleaning agent and as a refrigerant.

Effects of exposure

Inhalation of the vapour can produce drowsiness, giddiness, unconsciousness and death. If the fumes are drawn through the lighted tip of a cigarette, the danger is greatly increased due to the formation of acidic products, including phosgene. Its vapour may cause eye irritation and blistering of the skin. The effects of exposure can be cumulative. Repeated exposure, particularly where people handle the substance on a regular basis, can further result in dermatitis (eczema), particularly of the hands, due to the degreasing effect on the skin.

Control strategies

In addition to formal instructions to staff on the hazards of exposure and on the operation of formal safe working procedures, local exhaust ventilation should be provided around the lips of degreasing tanks (lip extraction). Trichloroethylene should be used in well-ventilated areas and its use in confined spaces prohibited.

Isocyanates

A range of isocyanates is used in industry, principally in the manufacture of urethane foams and resins, including toluene di-isocyanate (TDI), methylene bisphenyl di-isocyanate (MDI) and 1:5 naphthalene di-isocyanate (NDI).

Effects of exposure

TDI is an extremely volatile and toxic compound and the vapour can be produced during the manufacture of foams and hot wire cutting of the finished product.

Symptoms are usually reduced when the patient is removed from contact, but a severe respiratory reaction, as with MDI, may follow on second or subsequent exposures, even if the exposure is to extremely low concentrations of the vapour.

Most isocyanates can produce varying degrees of dermatitis and exposure can result in skin sensitisation in rare cases. The main consideration, however, is that isocyanates are a potent primary irritant to the respiratory tract and, in some cases, cause dramatic sensitisation i.e. asthmatic effects once the individual has become sensitised. Eye splashes may cause severe conjunctivitis.

Asthma due to exposure to isocyanates is a prescribed occupational disease.

Control strategies

The degree of volatility of a vapour affects the degree of inhalation and therefore the risk associated with a particular substance. Vapour pressure, namely the percentage gas by volume in air, is particularly important in controlling the hazards associated with isocyanates. Vapour pressure increases very rapidly with temperature.

Handling of isocyanates in open vessels should be prohibited and transference carried out through a fully enclosed system. Any spillages should be removed immediately and decontaminants should be readily available e.g. 5 per cent ammonia in sawdust.

Health surveillance should include pre-employment and routine periodic health examinations for all operators and other persons coming into contact with isocyanates. Training and instruction in the hazards and precautions necessary is vital.

CONTROL STRATEGIES FOR CARCINOGENS OR SUSPECTED CARCINOGENS

A carcinogen is a substance which is known to cause, or is suspected of causing, cancer. Well-established carcinogens include 2-naphthylamine, benzidine, vinyl chloride monomer (VCM), asbestos, coal tar pitch, mineral oils, hardwood dusts, arsenic and beryllium. Some carcinogens have a 10–40 year latent period. This is the time between the first exposure to the carcinogen and the actual diagnosis of the tumour.

Many carcinogenic hazards have been discovered over the last 75 years. The occupations affected, the agents causing cancer and the body sites affected are indicated in Table 2.1.

TABLE 2.1 Exposure to carcinogenic substances

Occupation	*Agent*	*Site*
Chimney sweeps Distillers of brown coal Makers of patent fuels Road workers, boat builders and others exposed to tar and pitch Cotton mule spinners	Combustion products of coal, shale oil (polycyclic hydrocarbons)	Scrotum and other parts of skin; bronchus
Dye manufacturers, rubber workers, manufacturers of coal gas	alpha- and beta-naphthylamine, benzidine and methylene bis-orthochloro-aniline (MBA or MBOCA)	Bladder
Radiologists, radiographers	Ionising radiation and X-rays	Skin
Chemical workers	4-aminodiphenol	Bladder
PVC manufacturers	Vinyl chloride monomer (VCM)	Liver
Haematite manufacturers	Radon	Bronchus
Asbestos workers, insultation workers, dock workers	Asbestos	Bronchus, pleura and peritoneum
Chromate workers	Chrome ore and pigments	Bronchus
Workers with glues and varnishes	Benzene	Marrow (myeloid and

Effects of exposure

Exposure to a wide range of substances can cause the development of a tumour or tumours in specific parts of the body or generalised throughout the body. A tumour or **neoplasm** (new growth) consists of a mass of cells which have undergone some fundamental and irreversible change in their physiology and structure which leads to a continuous and unrestrained proliferation of such cells.

The rate of growth of neoplasms varies greatly. Some may take many years to develop, whilst others may show an increase within a few days and extend far beyond their point of origin. This variation in growth behaviour and rate forms the basis of a broad classification of tumours into the benign and malignant classes.

Benign tumours are those which grow slowly and less expansively and, unless they occur in some vital site e.g. the brain, or interfere with an important organ, are well tolerated and do not necessarily interfere with a person's well-being or shorten his life. They are composed of well-differentiated and mature types of cells. The cells of the neoplasm resemble the cells of the original tissue.

Malignant tumours, on the other hand, grow more rapidly, will infiltrate and extend into normal tissue and structures and, unless effectively treated, interfere with health and eventually cause death. They are often composed of more embryonic (primitive) or poorly differentiated cell types, resembling less the cells of origin. They spread by means of secondary deposits or **metastases** to other sites in the body.

There are three ways in which tumours spread to form metastases: by the lymphatic system, the circulatory system or by transcoelomic spread, i.e. from the pleural or peritoneal cavities.

General control strategies

In most cases the handling, use and storage of known carcinogens is prevented by a legal prohibition under the COSHH Regulations. Schedule 2 of these regulations places a prohibition on certain substances hazardous to health for certain purposes. On this basis 2-naphthylamine, benzidine, 4-aminodiphenol, 4-nitrodiphenol their salts and any substance containing any of those compounds, and in any other substance in a total concentration exceeding 0.1 per cent, are prohibited in manufacture and use for all purposes including any manufacturing process in which any of these substances is formed.

In the case of other known or suspected carcinogens, such as auramine, magenta, VCM, orthotolidine, dianisidine and dichlorbenzidine, medical surveillance and the maintenance of health records is required for any person who may be exposed to risk.

Asbestos

Effects of exposure

Exposure to asbestos fibres in the workplace and in other environments may cause asbestosis. Approximately 50 per cent of asbestosis sufferers develop lung cancer and/or cancer of the bronchus.

Asbestosis is a fibrotic condition of the lung, resulting in scarring and thickening of the lung tissue. This condition may arise after many years of exposure to high concentrations of asbestos dust, although cases of asbestosis have been reported among people with only minimal exposure. The characteristic symptoms are a progressive breathlessness and unproductive cough. Lung damage takes the form of diffuse fibrosis or scarring throughout the lungs accompanied by emphysema and collagenous thickening of the pleural lining. The skin may have a bluish discolouration due to cyanosis and sputum may contain asbestos bodies. Other features include the presence of pleural placques (small patches attached to the pleura and peritoneum), asbestos bodies in the lung tissue and asbestos warts on the hands where the fibres penetrate the skin. A characteristic feature is also finger clubbing, a thickening of the fingers. Evidence so far suggests that asbestosis usually has a long induction or latent period of 10 to 20 years, and there is abundant evidence that those who smoke suffer both a greatly enhanced risk of contracting the disease and the prospect of a much worse overall lung condition.

Asbestosis, as a prescribed occupational disease, is described in three specified forms:

1 Diffuse mesothelioma (primary neoplasm of the mesothelium of the pleura or of the pericardium or of the peritoneum).
2 Primary carcinoma of the lung where there is accompanying evidence of one or both of the following:
   - asbestosis
   - bilateral diffuse pleural thickening.
3 Bilaterial diffuse pleural thickening; associated in all three cases with:
   - the working or handling of asbestos or any admixture of asbestos; or
   - the manufacture or repair of asbestos textiles or other articles containing or composed of asbestos; or
   - the cleaning of any machinery or plant used in any of the foregoing operations and of any chambers, fixtures and appliances for the collection of asbestos dust; or
   - substantial exposure to the dust arising from any of the foregoing operations.

Control strategies

Legal requirements relating to asbestos are covered by the Control of Asbestos at Work Regulations 1987 together with the ACOP 'Work with Asbestos Insulation and Asbestos Coating'. (See further Chapter 9.)

Control strategies for preventing exposure to asbestos include the following:

- identification by sampling and analysis of different forms of asbestos, or mixtures of same, which may be present as insulation or incorporated in specific structural finishes (see Fig 2.3)
- risk assessment prior to, particularly, stripping or removal of asbestos with a view to identifying potential exposure situations
- treatment of surfaces of insulation materials containing asbestos to seal or encapsulate same
- enclosure of the working area, particularly during dry stripping operations, together with the use of air extraction and filtration equipment

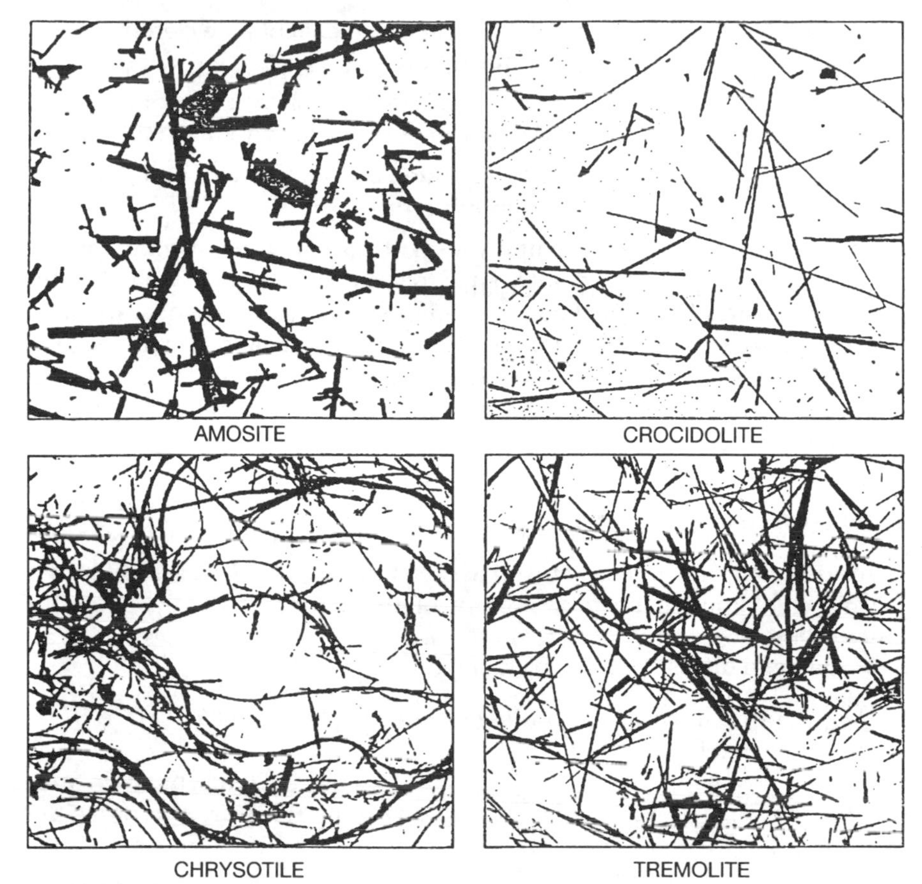

• **FIG 2.3 Electron micrographs of dispersed samples of the main types of asbestos fibres**

Source: Wagner and Elmes, 'The Mineral Fibre Problem' in *Recent Advances In Occupational Health*, Ed. J. C. McDonald, Churchill Livingstone, 1981

- regular examination of any enclosure together with air monitoring inside the enclosure
- the provision and use of personal 'approved respiratory protective equipment' by operators, provision of cleaning and maintenance arrangements for same together with separate accommodation for storing the equipment
- the provision and use of specified personal protective equipment, namely overalls, head coverings and footwear, together with effective arrangements for cleaning the protective clothing
- operation of clearly defined work procedures, established in a formally written Method Statement, with close supervision to ensure conformity with these procedures
- disposal of asbestos waste at licensed sites
- provision and use of separate purpose-built welfare amenity facilities incorporating showering facilities, separate storage arrangements for per-

sonal protective equipment/clothing and personal off-site clothing

- procedures for ensuring the maintenance and testing of plant and equipment – air extraction equipment, respiratory protective equipment, welfare facilities, vacuum cleaning equipment and the working enclosure
- the keeping of records relating to maintenance and repair of all plant and equipment, air monitoring and training provided
- training and supervision of employees with particular reference to the correct use of equipment and adherence to systems of work established, the need for workplace cleanliness and high standards of personal hygiene, and the need to report defects in equipment or inadequacies in the systems of work which may put people's health at risk. Training should take the form of both induction training for new employees and refresher training for existing employees.

Tar, pitch and bitumen

Exposure to tar, pitch, bitumen and associated substances can result in two specific conditions, namely dystrophy of the cornea of the eye and skin cancer. Both these conditions are prescribed occupational diseases which are defined thus:

1 Dystrophy of the cornea (including ulceration of the corneal surface) of the eye associated with:
   - the use or handling of, or exposure to, arsenic, tar, pitch, bitumen, mineral oil (including paraffin), soot or any compound, product or residue of any of these substances, except quinone or hydroquinone; or
   - exposure to quinone or hydroquinone during their manufacture.

2 Localised new growth of the skin, papillomatous or keratotic; and squamous-celled carcinoma of the skin, associated with:
   - the use or handling of, or exposure to arsenic, tar, pitch, bitumen, mineral oil (including paraffin), soot or any compound, product or residue of these substances, except quinone or hydroquinone.

Effects of exposure

All the substances referred to above are liable to be or contain irritants which, on continued contact, exert a damaging effect on mucous membranes and the skin. Such damage may be of, initially, an inflammatory nature. However, in the patent fuel industry, for instance, where anthracite dust and coal tar pitch are handled, inflammation and ulceration of corneal conjunctiva, slow to heal, may result in permanent scarring and interference with vision. Pitch workers are also known to suffer from acute dermatitis of exposed skin surfaces and may at the same time suffer from corneal complications.

People regularly exposed to these agents may also develop chronic skin lesions, such as localised new growths of a papillomatous or keratotic nature. These new growths, seen typically in the past in bituminous shale workers, commonly commence with a benign stage. If untreated, however, they may adopt an active and unregulated mode of growth. With people involved in extracting oil from shale, these growths commonly appear on the arms and other exposed parts of the body.

Where exposure to tar or pitch is in the form of fumes and dust, operators may also suffer from skin growths of a warty nature which, if untreated, tend later to ulcerate and assume a malignant character. Tar or pitch may thus initiate in an area of skin a neoplastic change while an individual is at work but malignant warts or progressive squamous-celled carcinoma may not appear until long after this exposure has ceased. This latent period is of varying duration.

Control strategies

These substances come within the general definition of a 'substance hazardous to health' in the COSHH Regulations 1994.

The provision and maintenance of a high standard of welfare facilities – showers, washing facilities, separate clothing storage arrangements – is essential in the prevention of skin cancer. This should be supported by the provision of information, instruction and training in the hazards and precautions necessary and in the need for high standards of personal hygiene.

Fumes and dust produced by tar, pitch or bitumen processing should be prevented or adequately controlled by the provision and maintenance of LEV systems, which should be subject to regular examination and testing. Respiratory and personal protective equipment, including eye protection, should be provided and its use enforced by management. Regular air monitoring of operating areas should be undertaken. Health surveillance should be provided for all exposed persons and continuing medical surveillance is required for persons employed in the manufacture of blocks of fuel consisting of coal, coal dust, coke or slurry with pitch as a binding substance.

Vinyl chloride monomer (VCM)

VCM as a gas is polymerised, when heated under pressure, to form polyvinyl chloride (PVC). Procedures under the COSHH Regulations for controlling the risks from VCM are outlined in the ACOP 'Control of Vinyl Chloride at Work'.

Effects of exposure

VCM is a highly toxic substance, and exposure, through inhalation, to VCM vapour in the industrial processing of VCM, results in absorption into the blood. This is then deposited in the liver, such deposition being associated with angiosarcoma, a tumour of the liver.

Angiosarcoma is a prescribed occupational disease and is associated with

1 work in or about machinery or apparatus used for the polymerisation of vinyl chloride monomer, a process which, for the purposes of this provision, comprises all operations up to and including the drying of the slurry

produced by the polymerisation and the packing of the dried product; or
2 work in a building or structure in which any part of that process takes place.

Symptoms include abdominal pain, reduced appetite, distension of the stomach, weight loss and jaundice. In many cases the disease is fatal.

Other diseases and conditions associated with exposure to VCM include fibrosis of the liver, impaired lung function, osteoporosis (loss of bone calcium), narcosis, caused by inhalation of high concentrations of VCM vapour, and marked changes in the skin, particularly hardening of same.

Control strategies

Wherever there is potential exposure to VCM, high levels of personal hygiene control are vital, supported by regular health surveillance of potentially exposed persons. Physical controls include well-designed local exhaust ventilation systems, well-controlled work systems, methods to reduce exposure time and correct use of PPE.

Methylene bis-orthochloroaniline, 2.2 dichloro-4.4 methylene dianiline (MBOCA or MBA)

MBOCA is produced mainly by the reaction of formaldehyde with o-chloroaniline and is commonly used in the manufacture of abrasion-resistant urethane rubbers and moulded semi-rigid polyurethane foam products. These intermediate products are incorporated in many final products, such as anti-vibration mounts, certain types of wheel and roller, and cable connectors and seals.

Pure MBOCA is a crystalline colourless solid, whereas commercial grades are a yellow-brown colour, and may contain varying quantities of polyamines and o-chloroaniline.

Effects of exposure

Generally, due to limited research into its toxicological properties, MBOCA is seen as carcinogenic to animals and gives positive tests as a mutagen for certain organisms. On this basis, the presumption is made that it is carcinogenic to humans. Absorption through the skin (pervasion) is seen as the principal route of entry.

Control strategies

This substance has been assigned a maximum exposure limit (MEL) of 0.005 mg/m (8-hour TWA). As a potential human carcinogen, prevention of exposure is, therefore, essential. As with many other potential carcinogens, high standards of personal hygiene are vital to prevent skin contact.

CONTROL STRATEGIES FOR CORROSIVE AND IRRITANT SUBSTANCES

Substances classified as 'corrosive' or 'irritant' are regulated by the various provisions of the COSHH Regulations. Corrosive substances include a wide range of acids and alkalis, such as sulphuric acid, hydrochloric acid, nitric acid, chromic acid, caustic soda and caustic potash, and ammonia. Irritant substances include both acids and alkalis, solvents and various metals and their salts, such as nickel and chromium, chromates and dichromates. Both classes of substance have varying effects on the skin and may affect certain target organs, such as the lungs. In certain cases, people may become sensitised following an initial exposure.

Chromium

Chromium is a hard, steel-grey, brittle metal. It forms two series of salts, the trivalent and the hexavalent, the latter being of great industrial importance, for instance, in chromium plating. It also forms an acid oxide, chromium trioxide (chromic acid), from which chromates and dichromates are formed.

Chromium ulceration is a prescribed occupational disease, described in the Social Security (Industrial Injuries) (Prescribed Diseases) Regulations 1985 (SS(II)(PD)R) thus:

Non-infective dermatitis of external origin (including chrome ulceration of the skin but excluding dermatitis due to ionising particles or electro-magnetic radiations other than radiant heat) associated with exposure to dust, liquid or vapour or any other external agent capable of irritating the skin (including friction or heat but excluding ionising particles or electro-magnetic radiations other than radiant heat).

The risk of contracting this form of occupational dermatitis is associated with the following uses of chromium:

- as metallic chromium in the formation of alloys
- as chromium compounds in chromium plating and anodising, in metal treatment processes, as tanning agents in the leather industry, in the impregnation of timber for preservation, as a constitutent of anti-corrosion paints, as sensitisers in the photographic industry and in the manufacture of dyestuffs.

Effects of exposure

Chromium comes within that group of external irritants responsible for occupational dermatitis. The common symptom, however, is the formation of chromium ulcers or chromium holes on the hands and forearms due to direct contact. Chromium plating processes produce chromic acid mist, exposure to which can result in chromium ulcers to the eyelids and, if this mist is inhaled, a hole can be formed right through the nasal septum. Chromates and dichromates used in, for instance, the manufacture of cement, may cause skin irritation and ulceration.

Chromium ulceration of the skin may be caused by exposure to chromic acid, the alkali chromates and dichromates, and zinc chromate. The penetration of these substances through an abrasion or minute break in the skin may cause a raised hard lump which breaks down at the centre revealing a deep ulcer with rounded and thickened edges and a slough-covered base.

Sensitisation, an allergic condition, is common in the case of people exposed to chrome salts. Once an individual has become sensitised, even the slightest exposure will produce a skin response.

Control strategies

Prevention and control strategies include the following:

- the provision, maintenance and regular testing and examination of LEV systems (lip extraction principally) to chromium plating processes
- the provision, maintenance and daily cleaning of impervious floor surfaces around plating baths
- the provision of impervious aprons, wellington boots and rubber gloves by employers and correct use of same by employees
- the provision and maintenance of welfare amenity provisions to a high standard, including provisions for the storage and drying of protective clothing
- pre-employment health screening of new employees and medical surveillance of employees, the maintenance of individual health records and frequent skin inspections by a responsible person, as required by the COSHH Regulations
- air monitoring of processing areas, as required by the COSHH Regulations, at least once every 14 days
- the enforcement of strict personal hygiene procedures, and the provision of barrier creams to be applied by employees prior to commencement of work and following washing of the hands and arms
- the provision of information, instruction and training for staff in the hazards and precautions necessary, including the need for regular skin examination and to report any form of skin irregularity to the occupational physician or occupational health nurse immediately.

Ammonia

Ammonia is a colourless gas with a highly pungent odour. It is commonly used as a refrigerant, in water treatment and in many processes, such as the manufacture of fertilisers, certain drugs, chemical preparations and in the refining of petroleum.

Effects of exposure

Ammonia burns, through splashing of the liquid on to the skin, can have serious and even fatal results. The effect of contact with the eyes can result in damage to the conjunctiva and cornea, including ulceration, tissue scarring, damage to vision and, in serious cases, blindness. Exposure through inhalation of the vapour results in tightness of the chest, and may be accompanied by smarting of the eyes, pain on swallowing, redness and swelling of the eyelids and conjunctiva. Continued exposure may lead to lung congestion and oedema.

Control strategies

Many items of equipment contain ammonia, such as ammonia compressors. These items should be subject to regular inspection and testing. It is common practice to install ammonia detectors in such areas which operate at well below the OEL. Respiratory protection must be provided by the use of gas masks or canister respirators which must be stored in a readily accessible location outside, for instance, ammonia compressor rooms. Safe systems of work should include the operation of permit to work systems when working on ammonia plant, supported by instruction and training of persons liable to come into contact with the vapour.

CONTROL STRATEGIES FOR SUBSTANCES CAUSING RESPIRATORY DISORDERS

Many respiratory disorders are associated with the inhalation of dusts, gases and other forms of airborne contaminant. These include the various forms of pneumoconiosis (coal worker's pneumoconiosis, asbestosis, silicosis and byssinosis) and carbon monoxide poisoning.

Pneumoconiosis

The group of lung disorders of a chronic fibrotic nature due to the inhalation of dust are generally classified as pneumoconiosis. This disorder is defined by the International Labour Organisation (ILO) as 'the accumulation of dust in the lungs and the tissue reaction to its presence'.

Pneumoconiosis is divided by the ILO into the **collagenous** and **non-collagenous** forms. (Collagen is a protein-based substance which forms the principal component of connective tissue. Its molecules are assembled like three-strand ropes.) The collagen diseases or connective tissue diseases have as their common factor a disorganisation of collagen strands. In all collagen diseases there is inflammation without infection.

Effects of exposure

Non-collagenous pneumoconiosis is caused by non-fibrogenic dust and has the following characteristics – intact alveolar architecture, minimal supporting tissue reaction and potentially reversible effects. Collagenous pneumoconiosis, on the other hand, may be caused by fibrogenic dusts or by an altered tissue response to a non-fibrogenic dust. It has the following characteristics – authenticated damage to alveolar architecture, appreciable supporting tissue reaction and permanent (irreversible) scarring of the lung.

Table 2.2 shows the types of collagenous pneumoconiosis that may result following inhalation of various dusts.

Table 2.2 The effects of dust exposure

Type of pneumoconiosis	*Dust*
Anthracosis	Coal dust
Silicosis	Free silica particles in gold, zinc, tin, iron and coal mining, sand blasting, metal grinding, slate quarrying, granite, sandstone and pottery work
Siderosis	Iron particles
Lithosis	Stone particles
Asbestosis	Asbestos fibres
Byssinosis	Cotton fibres

Control strategies

Effective methods for preventing the inhalation of dust are essential in the prevention of dust-borne diseases and disorders. In the selection of dust control measures a number of factors are important, in particular:

- the type of dust in terms of particle size, weight, density and air velocity
- the source of dust in a particular process
- the number of personnel exposed, the duration of exposure (continuous or intermittent) per day, and the number of days per week this emission takes place.

Methods of monitoring emissions, for instance by the use of static sampling equipment and/or personal dosimeters, and the results of past monitoring activities should be considered together with the system for the maintenance and examination of dust arrestment plant.

Implementation of all the following strategies is essential presuming, of course, that replacement or substitution of the offending agent with a less dangerous agent is not possible:

- suppression of the dust at source through the use of a wet process, as opposed to a dry process

- isolation, which entails enclosure of the complete process or the actual point of dust production as, for instance, in the case of tipping hoppers
- the operation of efficient LEV systems linked to collection and filtration plant, such as cyclone arrestors, dry deduster units, wet arrestors or electrostatic precipitators
- the maintenance of high standards of cleaning and housekeeping through the use of mechanical vacuum cleaning equipment, including *in situ* vacuum systems, as opposed to brushes and brooms
- medical surveillance of exposed personnel for early detection of respiratory conditions, supported by regular health examinations by occupational health nurses
- the supply, maintenance and use of personal protective equipment which implies the provision of the correct type of respiratory protective equipment according to the dust hazard involved, which the operator should use all the time that he is exposed to the dust, e.g. one-piece boiler suit, cap and gloves (see the requirements of the Personal Protective Equipment at Work Regulations 1992)
- the provision and maintenance of a high standard of welfare amenity provisions, in particular showering and separate workwear and personal clothing storage facilities
- the provision of information on the hazards and precautions necessary and the regular training of management and employees in the maintenance of the above procedures.

Coal worker's pneumoconiosis (CWP)

There are two forms of this disease, simple and complicated. Simple CWP is a relatively harmless condition, whereas complicated CWP (progressive massing fibrosis) is a different matter altogether.

Effects of exposure

With simple CWP there are nodular lesions in the lungs with little fibrosis. Emphysema, the abnormally distended condition of the lungs, is slight and there is little distortion of the lung architecture. This condition does not progress in the absence of further dust exposure. Neither, does it regress, however.

In the case of complicated CWP, dust collections are embedded in the diseased areas of fibrous tissue, and there is considerable distortion of the lung architecture and elasticity producing loss of and interference with the lung function.

There is still considerable argument as to whether simple CWP progresses to complicated CWP or whether they are two separate disease entities.

Control strategies

The measures described above apply in this case.

Silicosis

Silica takes a number of forms in both the crystalline and amorphous state, namely:

1 *crystalline* tridymite, cristobalite, quartz
2 *amorphous (after heating)* vitreous silica, diatomite, silica fumes and dried silica gel.

Potential sources of silicosis in industry are:

- *potteries, tile making* the drying out process produces silica dust
- *masonry* granite polishing, cutting, chipping and quarrying
- *furnaces* cutting of refractory bricks, stripping of furnaces
- *ceramics* manufacture of insulators
- *mining* coal-face working, coal washing
- *steel foundries* foundry sand, parting powders
- *sand-blasting processes* use of sandstone wheels.

Effects of exposure

Silicosis is a condition resulting in fibrosis of the lung. Nodular lesions are formed which ultimately destroy the lung structure. It is caused by the inhalation of particles of free silica within the respirable range. There is a strong predisposition to tuberculosis as a result of contracting silicosis.

Control strategies

The measures described above apply in this case.

Carbon monoxide

Carbon monoxide (CO) poisoning is a well-established occupational condition through exposure to coal gas, vehicle exhausts and most smokes.

Effects of exposure

CO combines with the haemoglobin of the blood in exactly the same way as oxygen, but it does so 300 times as readily as oxygen, and the victim dies of asphyxiation. It is characterised in the victim by a bright pink complexion and is treated by the administration of pure oxygen if possible, or otherwise fresh air, which gradually displaces the CO from the blood. In many cases, resuscitation may be needed.

The principal danger is that CO is colourless, odourless and tasteless and, therefore, in exposure situations goes undetected.

Initial signs include loss of balance and giddiness, tightness of the chest and loss of sensation and power in the legs, with the victim eventually falling unconscious.

Control strategies

CO poisoning arises in the classic gassing accident situation. Controls, therefore, hinge on recognition of potential situations and locations where CO could be unexpectedly produced in sufficient quantities to cause asphyxiation, the operation of permit to work systems in such situations, such as work in confined spaces, linked with the use of respiratory protection.

Similarly, high-risk areas should be well-ventilated by natural and/or mechanical ventilation. Ventilation of exhaust gases, including CO, by the use of flexible hoses linked to vehicle exhausts, for instance, during the running of vehicles, should be provided and used in enclosed vehicle servicing areas.

3

Physical health hazards

Physical stressors include extremes of temperature, lighting, ventilation and humidity, noise and vibration, radiation and work situations which require repetitive movements of part of the body, such as the hand or the arm.

The diseases and conditions associated with physical stressors are well-established, going back, in some cases, to before the turn of the century. Many of these diseases are prescribed for the purposes of industrial injuries benefit under social security legislation.

RADIATION

Radiation is a form of energy which is emitted by a wide variety of sources and appliances. All matter consists of elements, such as lead, oxygen, carbon and the basic unit of any element is the atom, which cannot be further sub-divided by chemical means. The atom is an arrangement of three types of particles, namely **protons**, which have unit mass and carry a positive electrical charge, **neutrons**, which have unit mass but carry no charge, and **electrons**, which have mass about 2000 times less than that of protons and neutrons and carry a negative charge.

The process of ionisation implies the charging of an atom or atoms to produce **ions**, that is charged atoms or groups of atoms. Where the number of electrons does not equal the number of protons, the atom has a net positive or negative charge and is said to be ionised. Thus if a neutral atom loses an electron, a positively charged ion will result. Ionisation is, therefore, the process of losing or gaining electrons and occurs in the course of many physical and chemical reactions.

It is important to distinguish between ionising radiation and non-ionising radiation. The former can produce chemical changes as a result of ionising molecules upon which it is incident. This leads to alteration in living cells and, in some cases, to certain biological effects. Non-ionising radiation, on the other hand, does not have this effect and is usually absorbed by the molecules upon which it is incident, with the result that the material will heat up, as in the case of microwaves.

Ionising radiations encountered in industrial and other processes are principally alpha, beta, gamma and X-rays, neutrons, Bremsstrahlung and cosmic. People can be irradiated by sources from outside the body or from radionuclides deposited in the body. Non-ionising radiation includes lasers, ultra-violet and infra-red radiation, microwaves and visible light.

Inflammation, ulceration or malignant disease of the skin or subcutaneous tissues or of the bones or blood, or blood dyscrasia, or cataract, due to electro-magnetic radiations (other than radiant heat) or to ionising particles is a prescribed occasional disease.

The various forms of ionising and non-ionising radiation are considered below.

Ionising radiation

Alpha particles

An alpha particle may be said to consist of two protons and two neutrons bound together. It is therefore heavy and double-charged. These are helium nuclei, that is, helium atoms that have lost their two orbiting electrons and are, therefore, positively charged. Though they are ejected from the nucleus of the radioactive substance with considerable energy, they are relatively large particles and are easily absorbed by matter.

Beta particles

These are electrons, not from the orbiting electrons of the atom but from within the nucleus. They are ejected with great speed and have a range of up to 15cm in air.

Gamma rays

These are very high-energy electromagnetic waves that are emitted at the same time as the alpha and beta particles. They are similar in nature to light but of very much shorter wavelength. They are very highly penetrating, being capable of passing through several centimetres of lead.

X-rays

X-rays are similar to gamma rays and are emitted from metals when bombarded with high-energy electrons. They are produced by changes in the energy state of planetary electrons. As with gamma rays, X-rays are a discreet quantity of energy, without mass or charge, that are propagated as a wave.

Neutrons

A neutron is an elementary particle with unit mass and no electric charge. The most powerful source of neutrons is the nuclear reactor.

Bremsstrahlung

These are weak X-rays produced by negative beta particles impinging on heavy materials.

Cosmic rays

These are fundamentally high-energy ionising radiations from outer space. They have a complex composition at the surface of the Earth.

Non-ionising radiation

Lasers

This term is derived from 'light amplification by stimulated emission of radiation – laser. Lasers are very high-energy beams of light.

Ultraviolet

This form of radiation is produced in arc welding and exposure may cause the condition known as 'arc eye'. Ultraviolet radiation also burns the skin and may cause cataracts and inflammation of the cornea.

Infra-red

Infra-red radiation is emitted by all hot bodies, particularly radiant fires. Long-term exposure to even low doses of infra-red radiation may damage the eyes, burning the lens and causing heat cataracts.

Microwaves

These are emitted at extremely high radio frequencies and can penetrate the body and cause internal heating of organs, in particular the testes and the eyes.

Visible light

Certain intense sources of visible light, such as that produced by the sun, arc lamps and electric welding units can damage the eyes and the skin.

Effects of exposure to ionising radiation

The effects of a dose of ionising radiation vary according to the type of exposure, for instance, whether the dose was a local one, affecting only a part of the body, or general, affecting the whole body. Furthermore the actual length of time of exposure determines the severity of the outcome of such exposure.

Local exposure is the most common form of exposure and may result in reddening of the skin with ulceration in serious cases. Where exposure is local and the dose small, but of long duration, loss of hair, atrophy and fibrosis of the skin can occur.

General exposure, on the other hand, can have a range of effects from mild nausea to severe illness, with vomiting, diarrhoea, collapse and eventual death. General exposure to small doses may result in chronic anaemia and leukaemia. The ovaries and testes are particularly vulnerable and there is evidence that exposure to radiation reduces fertility and causes sterility.

Apart from the risk of increased susceptibility to cancer, radiation can damage the genetic structure of reproductive cells, causing increases in the number of stillbirths and malformations.

The unit of biological dose of radiation is the **Sievert**. Specified dose limits are dealt with in the Ionising Radiations Regulations 1985 (see further Chapter 9).

Control strategies

A significant feature of radiological protection is the form taken by the radioactive substance. Sources may be sealed or unsealed.

With a **sealed source** the source is contained in such a way that the radioactive material cannot be released, for example, in X-ray equipment. The source of radiation can be a piece of radioactive material, such as cobalt, which is sealed in a container or held in another material which is not radioactive. It is usually solid and the container and any bonding material are regarded as the source.

Unsealed sources of radiation take many forms – gases, liquids and particulates. As they are unsealed, entry into the body is comparatively easy.

Radiological protection is based on three specific factors – **time, distance** and **shielding**. For instance, radiation workers may be protected by limiting the length of **time** or duration of exposure to certain predetermined limits. They can be protected by ensuring that they do not come within certain **distances** of radiation sources. This may be achieved by the operation of restricted areas, barriers and similar controls. In addition they may be **shielded** by the use of absorbing material, such as lead or concrete, between themselves and the source to reduce the level of radiation to below the maximum dose level. The quality and quantity (thickness) of shielding varies for the type of radiation and energy level and varies from no shielding through lightweight shielding, such as 1-cm thick Perspex, to heavy shielding in terms of centimetres of lead or metres of concrete. The principal objective is to ensure no one receives a harmful dose of radiation.

The following procedures are required for all sources of radiation:

- pre-employment and follow-up medical examinations
- the appointment of qualified persons (radiation protection advisers)
- the maintenance of individual dose records
- the provision of information, instruction and training for radiation workers, together with methods for maintaining hazard awareness
- continuous and spot check radiation dose monitoring, through the use of personal dosimeters and film badges for classified persons
- the use of warning notices, designated controlled areas and supervised areas
- strict adherence to maximum dose limits.

In addition, the following measures are necessary in the case of unsealed sources of radiation:

- the provision and use of appropriate personal protective equipment, including respiratory protection, goggles, one-piece overalls or chemical grade suits
- effective ventilation of working areas
- enclosure/containment of sources to prevent leakage
- the provision and maintenance of impervious working surfaces
- immaculate working techniques
- the use of remote control systems.

The central objective is the avoidance of radioactive contamination.

Legal requirements are covered by the Ionising Radiations Regulations 1985 (see further Chapter 9).

NOISE

'Sound' is defined as 'any pressure variation in air, water or some other medium that the human ear can detect.' 'Noise', on the other hand, is generally defined as 'unwanted sound'. Noise can be a nuisance at common law and under statute law, resulting in disturbance and loss of enjoyment of life, loss of sleep and fatigue. Secondly, it can distract attention and concentration, mask audible warning signals or interfere with work, thereby becoming a causative factor in accidents. Finally, exposure to excessive noise can result in hearing impairment, the condition known as 'noise-induced hearing loss' or 'occupational deafness'. However, provided the intensity and duration of exposure are sufficient, even 'wanted sound', such as loud music, can lead to hearing impairment.

Occupational deafness is a prescribed occupational disease which is described thus (SS(II)(PD)R):

> *Substantial sensorineural hearing loss amounting to at least 50 dB in each ear, being due in the case of at least one ear to occupational noise, and being the average of pure tone loss measured by audiometry over the 1, 2 and 3 KHz frequencies (occupational deafness) associated with:*
>
> *(a) the use of, or work wholly or mainly in the immediate vicinity of, pneumatic percussive tools or high-speed grinding tools, in the cleaning, dressing or finishing of cast metal or of ingots, billets or blooms;*
>
> *or*
>
> *any occupation involving:*
>
> *(b) the use of, or work wholly or mainly in the immediate vicinity of, pneumatic percussive tools on metal in the shipbuilding or ship repairing industries; or*
>
> *(c) the use of, or work in the immediate vicinity of, pneumatic percussive tools on metal, or for drilling rock in quarries or underground, or in mining coal, for at least an aver-*

age of one hour per working day; or

(d) *work wholly or mainly in the immediate vicinity of drop-forging plant (including plant for drop-stamping or drop-hammering) or forging press plant engaged in the shaping of metal; or*

(e) *work wholly or mainly in rooms or sheds where there are machines engaged in weaving man-made or natural (including mineral) fibres or in the bulking up of fibres in textile manufacturing; or*

(f) *the use of, or work wholly or mainly in the immediate vicinity of machines engaged in cutting, shaping or cleaning metal nails; or*

(g) *the use of, or work wholly or mainly in the immediate vicinity of, plasma spray guns engaged in the deposition of metal; or*

(h) *the use of, or work wholly or mainly in the immediate vicinity of, any of the following machines engaged in the working of wood or material composed partly of wood, that is to say; multi-cutter moulding machines, planing machines, automatic or semi-automatic lathes, multiple cross-cut machines, automatic shaping machines, double-end tenoning machines, vertical spindle moulding machines (including high-speed routing machines), edge banding machines, bandsawing machines with a blade width of not less than 73mm and circular sawing machines in the operation of which the blade is moved towards the material being cut; or*

(i) *the use of chain saws in forestry.*

Effects of exposure

Noise may affect hearing in three ways:

1 **Temporary threshold shift** is the short-term effect, that is, a temporary reduction in hearing acuity, which may follow exposure to noise. The condition is reversible and the effect depends to some extent on individual susceptibility.
2 **Permanent threshold shift** takes place when the limit of tolerance is exceeded in terms of time, the level of noise and individual susceptibility. Recovery from permanent threshold shift will not proceed to completion, but will effectively cease at some particular point in time after the end of the exposure.
3 **Acoustic trauma** is quite a different condition from occupational deafness (noise-induced hearing loss). It involves sudden aural damage resulting from short-term intense exposure or even from one single exposure. Explosive pressure rises are often responsible, such as exposure to gunfire, major explosions or even fireworks.

For most steady types of industrial noise, intensity and duration of exposure (dose) are the principal factors in the degree of noise-induced hearing loss. Hearing ability also deteriorates with age (presbycusis), and it is sometimes difficult to distinguish between the effects of noise and normal age deterioration in hearing. Research by the UK Medical Research Council and the National Physical Laboratory has shown that the risk of noise-induced hearing loss can be related to the total amount of noise energy that is taken in by the ears over a working lifetime.

Symptoms of noise-induced hearing loss vary according to whether the hearing loss is mild or severe. Typical symptoms associated with a mild form of hearing loss include a difficulty in conversing with people, the wrong answers may be given occasionally due to the individual missing certain key elements of the question, and speech on television and radio seems indistinct. Moreover, there is difficulty in hearing normal domestic sounds, such as a clock ticking.

With a severe degree of deafness there is difficulty in discussion, even when face-to-face with people, as well as hearing what is said at public meetings, unless sitting right at the front. Generally, people seem to be speaking indistinctly, even on radio and television, and there is an inability to hear the normal sounds of the home and street. It is often impossible to tell the direction from which a source of noise is coming, and to assess the distance from the sound. In the most severe cases, there is a sensation of whistling or ringing in the ears (tinnitus).

Control strategies

In any strategy to reduce or control noise two factors must be considered, that is, the actual source of the noise, and the transmission pathway taken by the noise to the recipient. Personal protective equipment, such as ear plugs, ear defenders and acoustic wool, may go some way towards preventing people from going deaf at work, but such a strategy must be regarded as secondary since it relies too heavily upon exposed persons wearing potentially uncomfortable and inconvenient protection for all the time they are exposed to noise. The majority of people simply will or do not do this!

The first consideration must be that of tackling a potential noise problem at the design stage of new projects, rather than endeavouring to control noise once the machinery or noise-emitting item of plant is installed. Manufacturers and suppliers of machinery and plant must be required to give an indication of anticipated sound pressure levels emitted by their equipment and of the measures necessary, in certain cases, to reduce such noise emission, prior to the ordering of same.

In the case of existing machinery and plant, different methods of noise control are suitable for dealing with different sources and for the possible stages in the transmission pathway. These are summarised in Table 3.1.

Control of the main or primary pathway is the most important factor in noise control and to ensure compliance with the Noise at Work Regulations 1989. HSE Noise Guide No. 4, 'Engineering Control of Noise', produced in conjunction with the above regulations, makes the following recommendations:

1 Controlling the noise generated

Some of the methods of preventing machinery noise generation that should be considered are:

- avoiding impacts, or providing arrangements to cushion them, for example buffers on stops, the use of rubber or plastic surface coatings to avoid metal to metal impacts on chutes

LIBRARY
TECHNICAL COLLEGE
219377

TABLE 3.1 Methods of noise control

Sources and pathways	*Control measures*
Vibration produced through machinery operation	Reduction at source e.g. substitution with nylon components for metal; use of tapered tools on power presses
Structure-borne noise (vibration)	Vibration isolation e.g. use of resilient mounts and connections, anti-vibration mounts
Radiation of structural vibration	Vibration damping to prevent resonance
Turbulence created by air or gas flow	Reduction at source or the use of silencers
Airborne noise pathway	Noise insulation – reflection; use of heavy barriers Noise absorption – no reflection; use of porous lightweight barriers

- increasing damping to reduce the tendency of machine parts to 'ring', including treatments applied to sheet metals such as surface coatings
- installation of silencers to reduce noise generated by turbulence at air exhausts and jets, such as the use of a porous silencer for the exhaust of a pneumatic cylinder
- use of low-noise air nozzles, pneumatic ejectors and cleaning guns constructed on good aerodynamic principles, or substitution of an alternative method of doing the job, for example the use of a mechanical instead of pneumatic ejector on a power press
- matching of air supply pressure to the actual needs of air-powered equip ment by providing each unit with its own pressure-reducing valve
- arrangements to make sure that noisy devices are only used when actually needed
- improved design of fans, fan casings and compressors and their accurate matching to the systems they supply
- reducing the need for noisy assembly practices by better quality control, design and manufacturing procedures
- the use of flexible elements to reduce the spread of structure-borne sound through a machine frame, for example, isolating mountings for bearings to reduce transmission of gear noise to the gearbox casing
- dynamic balancing of rotating parts
- attention to the stiffness of structural parts of machines, for example

ensuring that if a C-frame power press is intended to be used with tie bars, they are in fact fitted

- noisy process elimination by better technology, for example better plate-cutting techniques to eliminate noisy rectification processes such as caulking with chipping hammers
- reducing noise levels by better machine maintenance
- avoiding 'chopping' of air streams by rotating components close to fixed parts of machines, such as a square cutter block rotating close to a fixed table or a fan blade passing close to a projecting part of the casing. Cutter blocks with helical casings can help reduce this noise, as will increasing the distance between the fixed and rotating components.

2 Substitution of a quieter machine or process

Sometimes a change of technology or the operation of a different working procedure can reduce noise.

3 Modification of the routes by which noise reaches workplaces

The path between points at which noise is generated and the workplace can sometimes be modified. Some of the measures which should be considered are:

- use of sound-absorbing material to control reflections within workrooms
- use of mufflers or silencers to reduce noise transmitted along pipes and ducts, for example on the exhaust and intake to internal combustion engines and on ducts to control noise from exhaust ventilation fans
- the installation of anti-vibration mountings under machines
- enclosure of noisy machines and partial enclosures or covers around noisy parts of machines (see Fig 3.1)
- enclosure of the workplace by provision of a cabin or noise refuge
- the use of screens faced with sound-absorbing material placed between working areas and local noise sources.

4 Distance

Increasing the distance between a person and the noise source can provide considerable improvements. Some ways of achieving this are:

- arrangements for exhausts to be discharged well away from workers
- the use of remote control or automated equipment to avoid the need for workers to spend long periods near to machines
- the segregation of noisy processes to restrict the number of persons exposed to high noise levels.

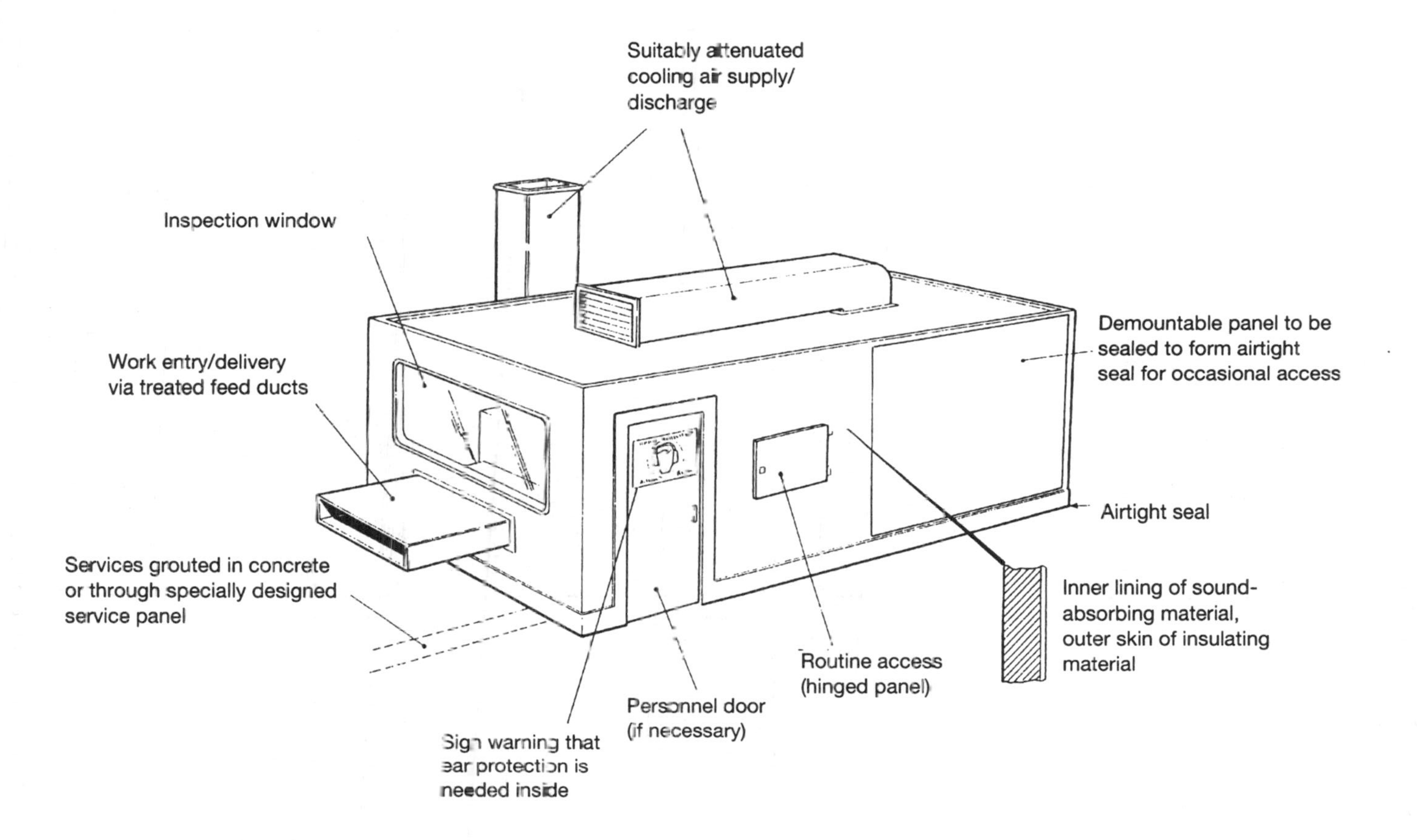

● **FIG 3.1 Typical machine enclosure**

Noise control programmes

A typical structure for a noise control programme is shown in Fig 3.2 below.

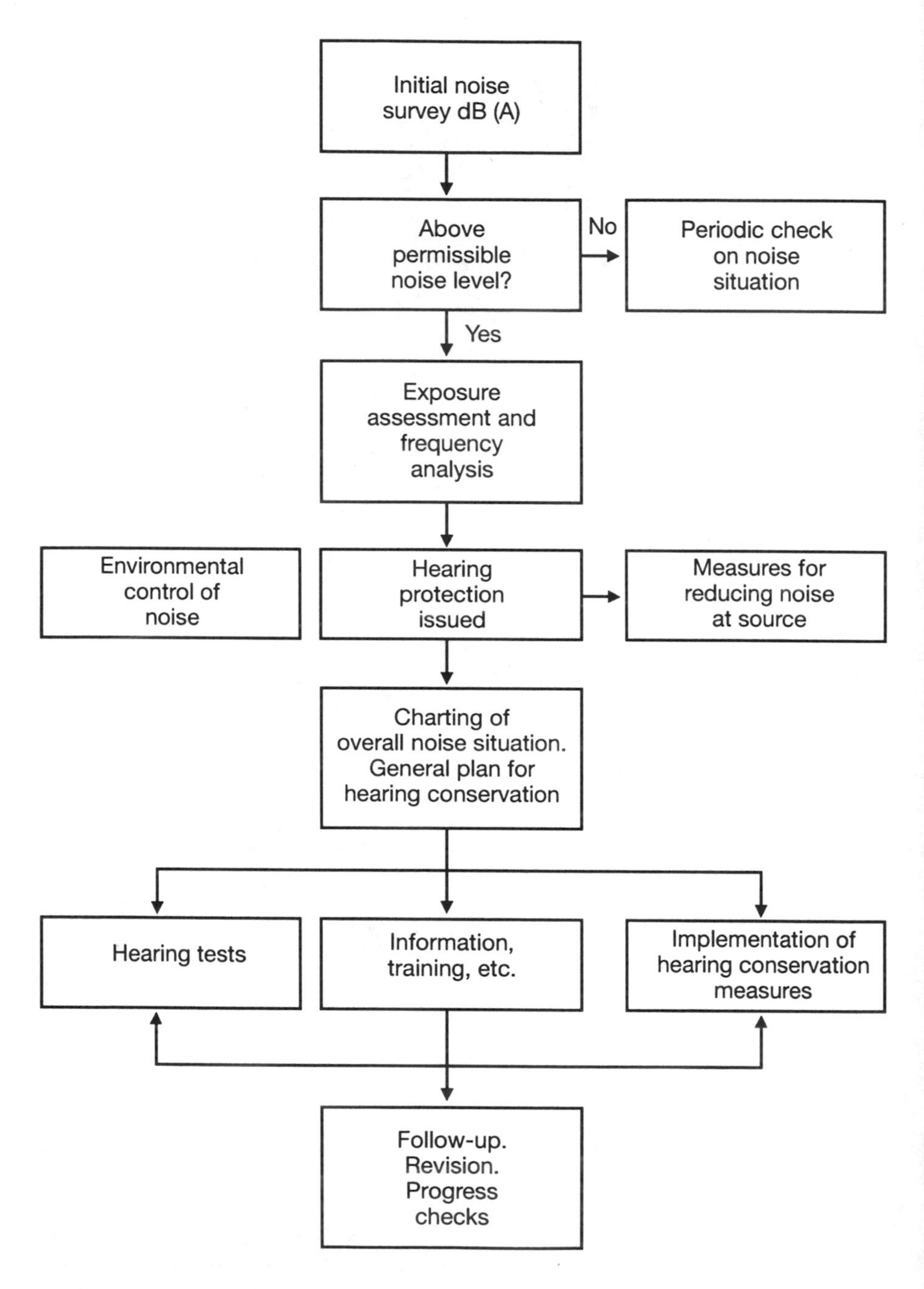

• **FIG 3.2 Typical structure for a noise control programme**

VIBRATION

The principal health condition related to vibration exposure is 'vibration-induced white finger' (VWF), which is associated with the use of vibratory hand tools, such as compressed air pneumatic hammers, electrically operated rotary tools and chain saws. Prolonged exposure to local vibration causes this condition. VWF is a prescribed occupational disease. People suffering from the similar condition, Raynaud's syndrome, are likely to have the condition dramatically worsened by exposure to vibration.

Vibration-induced white finger

The condition is defined as episodic blanching, occurring throughout the year, affecting the middle or proximal phalanges or in the case of a thumb the proximal phalanx, or:

- in the case of a person with five fingers (including thumb) on one hand, any three of those fingers
- in the case of a person with only four such fingers, any two of those fingers
- in the case of a person with fewer than four such fingers, any one of those fingers or ... the one remaining finger. (S S (I I) (P D) R 1985, schedule 4.)

3

Raynaud's syndrome

This may be caused by a number of conditions unassociated with exposure to vibration. In fact, research indicates that vibration is not the primary cause. It was originally seen as a manifestation of Raynaud's disease (or 'constitutional white finger') which is thought to be of a hereditary nature. The symptoms may also be caused by diseases of the connective tissue, such as rheumatoid arthritis, trauma, thrombotic conditions, blood disorders and certain neurological diseases. In order to put a specific label on a characteristic set of symptoms caused by exposure to vibration, the Industrial Injuries Advisory Committee has favoured the term 'vibration-induced white finger'.

Symptoms of VWF

The first signs of VWF, which often pass unnoticed or are not attributed to vibration, are mild tingling and numbness of the fingers, similar to 'pins and needles'. Later, the tips of the fingers which are most exposed to vibration become blanched, typically early in the morning or in cold weather. On further exposure, the affected area increases, sometimes to the base of the fingers, sensitivity during attacks is reduced and the characteristic reddening of the areas affected marks the end of an attack causing severe pain. Prolonged and intense exposure may cause further advancement of the condition with the fingers sometimes taking on a blue-black appearance. In severe cases, gangrene and necrosis (death of living tissue) have been reported. Table 3.2 shows a classification of the severity of VWF developed by Taylor and Pelmear in 1975.

There is a latent or symptom-free period from the commencement of regular exposure to the onset of ill effects, the length of which is thought to be related to the intensity of the vibration. In people subjected to very high intensity levels, stages 2 and 3 may be reached within a few months. Usually the condition progresses more slowly, and a typical latent period is around five years. In general, the shorter the latent period, the more severe the condition will become as exposure increases.

TABLE 3.2 Stages of vibration-induced white finger

Stage	*Condition of digits*	*Work and social interference*
0	No blanching of digits	No complaints
0_T	Intermittent tingling	No interference with activities
0_N	Intermittent numbness	No interference with activities
1	Blanching of one or more fingertip with or without tingling and numbness	No interference with activities
2	Blanching of one or more fingers with numbness; usually confined to winter	Slight interference with home and social activities; no interference at work
3	Extensive blanching; frequent episodes, summer as well as winter; restriction of hobbies	Definite interference at work, at home and with social activities
4	Extensive blanching; most fingers; frequent episodes, summer and winter	Occupation changed to avoid further vibration exposures because of severity of signs and symptoms

Note: Complications are not considered in this grading.
Source: Taylor and Pelmear, 1975

Other vibration-related conditions

These include:

1 **Osteoarthritis of the arm joints** this condition is encountered most commonly in the elbow joints, but also in the wrist and shoulder joints.
2 **Injury to soft tissues of the hand** this is mainly injury to the palm of the hand (Dupuytren's contracture) and bursitis; atrophy (shrinking or wasting).
3 **Decalcification of the carpus** this condition in the main bone at the base of the hand has been noted in the hands of workers using pneumatic tools, but it does not deteriorate.

Vibratory hand tools

With hand tools the energy level at particular frequencies is significant in the prevention of VWF. Percussive action tools in the range 2000–3000 'beats per minute', equal to a frequency range of 33–50 Hz, are the worst. With rotary tools, the range 40–125 Hz is common and promotes similar damage.

Many vibratory hand tools are used in industry, including pneumatic tools (for riveting, caulking, fettling, rock drilling and hammering), combustion engine-operated tools (for chain sawing, drilling, vehicle operation, the operation of flexi-driven machinery) and electrically powered tools (for grinding, concrete levelling, swaging, drilling and burring).

Reduction of vibration injuries

Vibration injury is best reduced either by redesign of the tool or by introducing more automation to isolate the operator's hand from the source of vibration. Alternatively, it may be possible to introduce a shock-absorbing mechanism between the vibration source and the handles of the tool. If this is impracticable, the following procedures should be carried out:

- pre-employment health screening of potentially exposed workers to assess susceptibility to VWF or manifestation of Raynaud's syndrome
- ensuring body warmth before the start of work in order to achieve good circulation to the extremities
- ensuring that the temperature in the workplace maintains body warmth throughout the working day
- in the case of outside workers, the supply and use of wind-resistant clothing and gloves together with replacements for wet clothing
- minimising smoking because of its effects on circulation
- where workers are reaching or have already reached the irreversible stage, i.e. stage 4, preclusion from further exposure
- regular health examinations
- redesign of portable hand tools at reduced frequencies
- mechanisation of grinding methods
- provision and use of cotton gloves with rubber inserts or padded with absorbent material

- use of specific working methods to reduce exposure time, such as job rotation
- proper maintenance of tools, such as sharpening of cutters, tuning of engines and renewal of vibration isolators
- training in correct working techniques to minimise exposure.

Whole body vibration

Whilst VWF tends to be the principal problem associated with vibration, the effects of vibration on the body generally should not be ignored. Whole body vibration associated with, for instance, driving heavy lorries long distances, can cause blurred vision, loss of balance and loss of concentration, the latter being a causative factor in road accidents.

In 1974 the International Standards Organisation published recommendations concerned with vibration and the human body (ISO 2631–1974). The recommendations cover cases where the human body is subjected to vibration on one of the three supporting surfaces i.e. the feet of a standing person, the buttocks of a person while sitting down and the areas supporting a lying person. Three severity criteria are specified:

- a boundary of reduced comfort, applying to fields such as passenger transportation
- a boundary of fatigue-decreased efficiency that is relevant to drivers of vehicles and to certain machine operators
- an exposure limit boundary, which indicates danger to health.

Research shows that in the longitudinal direction i.e. head to feet, the human body is most sensitive to vibration in the frequency range 4–8 Hz, whilst in the transverse direction i.e. fingertip to fingertip, the body is most sensitive in the range 1–2 Hz.

WORK-RELATED UPPER LIMB DISORDERS

Work-related upper limb disorders caused by repetitive strain injuries (RSI) were first defined in the medical literature by Bernardo Ramazzini, the Italian father of occupational medicine, in the early eighteenth century. The International Labour Organisation recognised RSI as an occupational disease in 1960 as a condition caused by forceful, frequent, twisting and repetitive movements.

RSI covers some well-known conditions such as tennis elbow, flexor tenosynovitis and carpal tunnel syndrome. It is usually caused or aggravated by work, and is associated with repetitive and over-forceful movement, excessive workloads, inadequate rest periods and sustained or constrained postures, resulting in pain or soreness due to the inflammatory conditions of muscles and the synovial lining of the tendon sheath. Present approaches to treatment are largely effective. Tenosynovitis has been a prescribed industrial disease since 1975, and the HSE has proposed changing the name of the condition to 'work-related upper limb disorder' on the grounds that the disorder does not always result from repetition or strain, and is not always a visible injury.

Many people, including assembly workers, supermarket checkout assistants and keyboard operators, are affected by RSI at some point in their lives. In January 1994, Kathleen Harris, who had worked for the Inland Revenue for 14 years, was awarded damages of £79,000 after contracting RSI.

Clinical signs and symptoms

These include local aching pain, tenderness, swelling and crepitus (a grating sensation in the joint) aggravated by pressure or movement. Tenosynovitis, affecting the hand or forearm is the second most common prescribed industrial disease, the most common being dermatitis. True tenosynovitis, where inflammation of the synovial lining of the tendon sheath is evident, is rare and potentially serious. The more common and benign form is peritendinitis crepitans, which is associated with inflammation of the muscle-tendon joint that often extends well into the muscle.

Forms of RSI

(See Figure 3.3.)

1 **Epicondylitis** inflammation of the area where a muscle joins a bone.
2 **Peritendinitis** inflammation of the area where a tendon joins a muscle.
3 **Carpal tunnel syndrome** a painful condition in the area where nerves pass through the carpal bone in the hand.
4 **Tenosynovitis** inflammation of the synovial lining of the tendon sheath.
5 **Tendinitis** inflammation of the tendons, particularly in the fingers.
6 **Dupuytren's contracture** a condition affecting the palm of the hand, where it is impossible to straighten the hand and fingers.
7 **Writer's cramp** cramps in the hand, forearm and fingers.

Prevention of RSI

Injury can be prevented by:

- improved design of working areas e.g. position of keyboard and display screens, heights of workbenches and chairs
- adjustment of workloads and rest periods
- provision of special tools
- health surveillance aimed at detecting early stages of the disorder
- better training and supervision.

If untreated, RSI can be seriously disabling.

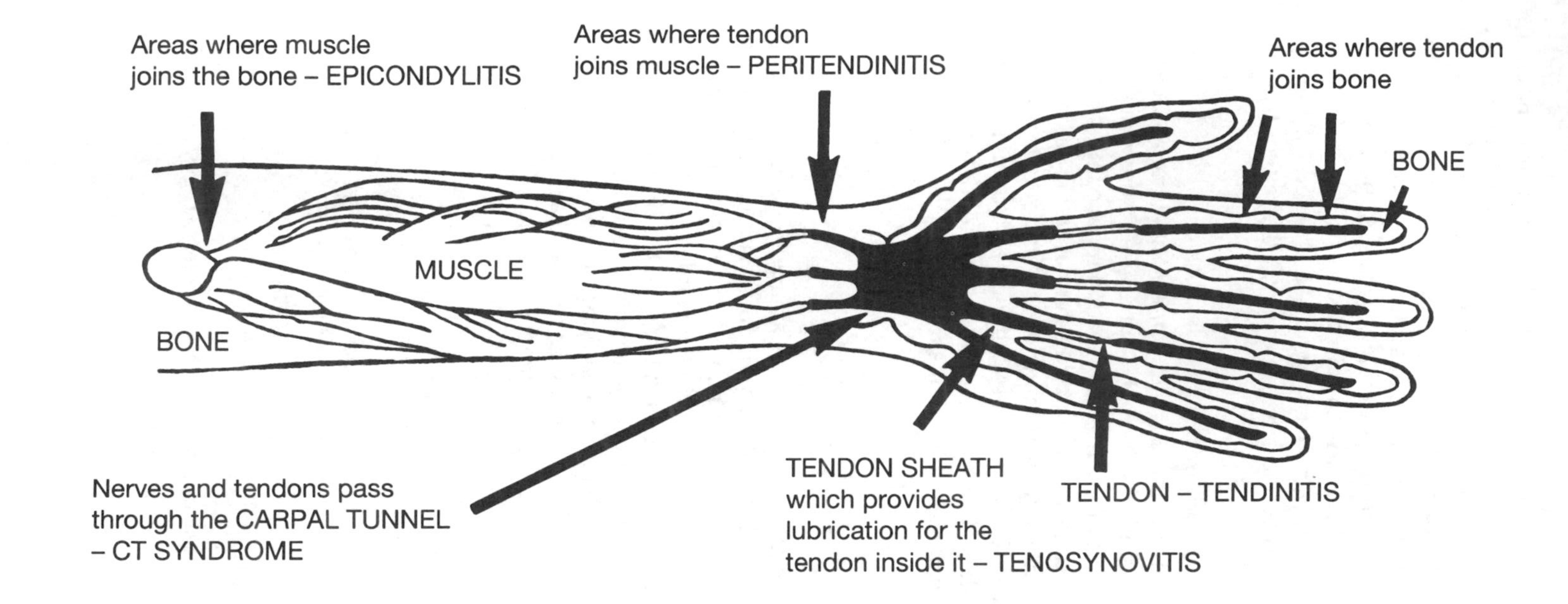

● **FIG 3.3. RSI – Where you find it**

THE BEAT DISORDERS

These are a group of occupational conditions affecting the hand, elbow and knee due to friction and pressure on limbs and joints.

Beat hand

Referred to as 'sub-cutaneous cellulitis of the hand', this condition or disability, prescribed disease A5 under the SS(II)(PD)R, is the result primarily of the bruising of the skin and the underlying tissues and the implantation there, by friction or pressure, of 'dirt' and particles. The condition is liable to follow frequent jarring of the hand in the use of a pick and shovel, and is more likely to occur in wet conditions. It is principally found in the hands of people unaccustomed to manual labour or who have been away from such activity for a long time. When accompanied by local infection, it may become acutely disabling. The condition is encountered in the palm of the hand and takes the form of, firstly, an acute inflammation, followed in many cases by a suppurative condition i.e. broken skin and the presence of pus, due to infection.

Beat knee

Officially described as 'bursitis or subcutaneous cellulitis arising at or about the knee due to severe or prolonged external friction or pressure at or about the knee', this condition, prescribed occupational disease A6, is similar in aetiology to beat hand. It occurs in those unaccustomed to working in a kneeling position or on returning to such work after a prolonged absence, and is more likely to occur if the skin is wet and sodden. Repeated or lengthy pressure, together with regular pivoting on the knee, as in the case of roof tilers or carpet fitters who persistently kneel, is a potential cause.

Cellulitis of the skin generally proceeds to the suppuration stage and may involve the bursa of the knee. In bursitis, the enlargement of the knee joint may be due to acute effusion (leakage of fluid into a body cavity) or to infection of a chronic enlargement. Depending upon the severity of the condition, incapacity may last only a few weeks or surgery may be necessary to remedy the condition.

Beat elbow

This condition is similar in aetiology to beat hand and beat knee, but with the elbow a single, although perhaps sustained, injury during work is more easily identified as the cause. Here again there are the classical signs of acute inflammation. The elbow is swollen and painful, signs of deep inflammation set in, and the swelling rapidly extends down the back of the forearm. The prognosis, as with other beat conditions, depends upon the degree of severity of the condition.

TEMPERATURE

A number of occupational conditions are associated with exposure to, particularly, high temperatures. These include heat cataract, heat stroke and heat cramps.

Heat cataract

Cataracts of the eye, caused by excessive exposure to heat and microwaves, have been common in many industries e.g. glass blowing, chain making and others requiring the operation of furnaces. Continuous exposure to radiant heat results in the opacity of the lens of the eye. Such radiations, it is thought, disturb the nutrition of the lens and cause localised coagulation of the protein. Heat cataract is included in the list of disorders under prescribed disease A2 under the SS(II)(PD)R.

Heat stroke

This condition is occasionally encountered in workers in hot processes. The symptoms are due to a defect in thermoregulation – the ability of the body to vary its temperature according to external factors. The onset is usually abrupt; the patient falls unconscious and could have a temperature of 40.6°C or more. Emergency treatment is aimed at reducing the body temperature to 40.0°C within one hour by all possible means, thus minimising the risk of damage to the central nervous system. Heat stroke is included in the list of disorders under prescribed disease A2 under the SS(II)(PD)R.

Heat cramps

Heat cramps may be encountered by workers engaged in the heat treatment of metals e.g. forging or casting, as a result of heat from microwave radiation. Most cases occur during summer months. Cramps take the form of pain in the muscles beginning in the calves and spreading to the arms and abdomen. The pains are of an intermittent nature, occurring with increasing severity every few minutes. Generally, taking a drink containing common salt (saline) is the only treatment necessary, and it is normal in many industries to have stocks of salt tablets available.

LIGHTING

Headaches, vertigo, insomnia and visual fatigue are common in many work situations. They are associated with poor lighting and the level of visual performance of the operator.

There is one occupational disease prescribed in relation to lighting (or the lack of it) namely miner's nystagmus, prescribed disease A9.

Miner's nystagmus

This condition is associated with poor lighting conditions in underground working operations, and is a complex psychological malady. It is thought to be primarily the result of poor lighting. However, exposure to toxic gas, the adoption of unusually awkward working postures or the onset of a state of anxiety have often been precipitating factors. The disease is associated with the more or less rhythmic oscillation of the eyeballs often coupled with persistent headaches, vertigo and insomnia. It may be accompanied by contraction of the fields of vision, poor visual acuity and photophobia, nervous symptoms and tremor, as well as nuchal rigidity (a condition of the brain) with a characteristic posture in walking. Generally, complete recovery takes place after cessation of underground work and a return to exposure to good illuminance levels.

ELECTRICITY

Effects of exposure

Electric shock is a possible outcome of electric current flowing through the human body, which causes disturbance in the normal functions of the body's organs and nervous system. Death occurs if the rhythm of the heart is upset for long enough to stop the flow of blood to the brain (ventricular fibrillation).

If a person is in contact with a live conductor, a material that readily conducts electricity, and another part of his body is touching a conducting path, such as an earthed metal pipe, then the voltage to earth of that conductor will cause current to flow, through the body's resistance, to earth. The amount of current flowing will depend upon the voltage, which is usually the standard 240 volts supply, and upon the resistance of the body and other parts of the conducting path for the current to earth.

For a given current flow through the body, the severity of electric shock depends upon the length of time that the current flows, but it must be realised that only a small current flowing for a moment in time can be dangerous. The effect of electric shock varies with age, sex, medical and physical condition and the body's resistance to current flow.

Most electrical injuries, however, arise from burns received at the point of contact with the body. However, some of these burns can be deep-seated and immediate medical treatment is essential.

Control strategies

Legislation covering electrical safety is the Electricity at Work Regulations 1989. These regulations are accompanied by the HSE's Memorandum of Guidance which provides excellent advice on electrical safety procedures.

The principal objective is that of protecting people from electric shock, and also from burns arising from contact with electricity and the effects of fire which may arise from defective electrical installations. There are two basic preventive measures against electric shock:

- protection against direct contact, for example, by providing proper insulation for parts of equipment liable to be charged with electricity
- protection against indirect contact, for example, by providing effective earthing for metallic enclosures which are liable to be charged with electricity if the basic insulation fails for any reason.

When it is not possible to provide adequate insulation against indirect contact, a range of measures is available, including protection by barriers or enclosures, or protection by position, that is placing live parts out of reach.

Principal protection is by means of earthing and the use of reduced voltage systems, details of which are outlined below.

Earthing

The provision of effective earthing, to give protection against indirect contact, can be achieved in a number of ways, including connecting the extraneous conductive parts of premises (water pipes, taps, radiators) to the main earthing terminal of the electrical installation. This would create an 'equipotential' zone and eliminate the risk of shock that could occur if a person touched two different parts of the metalwork liable to be charged, under earth fault conditions, at different voltages. It is crucial to ensure that in the event of earth fault, such as when a live part touches an enclosed conductive part (usually metalwork), that the electricity supply is automatically disconnected. Such disconnection is achieved by the use of overcurrent devices (correctly rated fuses or circuit breakers) or by correctly placed and rated residual current devices.

Reduced voltage

Another protective measure against electric shock is the use of reduced voltage systems, the most commonly used being the 110 volt centre point earthed system, that is, the secondary winding of the transformer providing the 110-volt supply is centre tapped to earth, thus ensuring that at no part of the 110-volt circuit can the voltage exceed 55 volts. Safe extra-low voltage systems are also available, which operate at 50 volts, but have limited though safer application.

Electric shock – emergency procedure

It is vital that operators be trained in the emergency procedure in the event of electric shock. The resuscitation procedure advocated by RoSPA is shown in Fig 3.4.

1 RECOGNISE A LACK OF OXYGEN

Arising from

ELECTRIC SHOCK
DROWNING
POISONING
HEAD INJURY
GASSING etc

May be causing

UNCONSCIOUSNESS
NOISY OR
NO BREATHING
ABNORMAL COLOUR

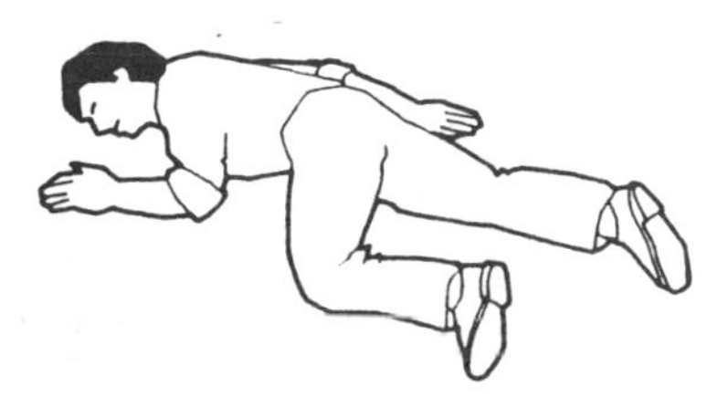

2 ACT AT ONCE

SWITCH OFF ELECTRICITY, GAS, etc.,
REMOVE CASUALTY FROM DANGER
SEND SOMEBODY FOR HELP

GET A CLEAR AIRWAY ...
REMOVE ANY OBSTRUCTION ... then

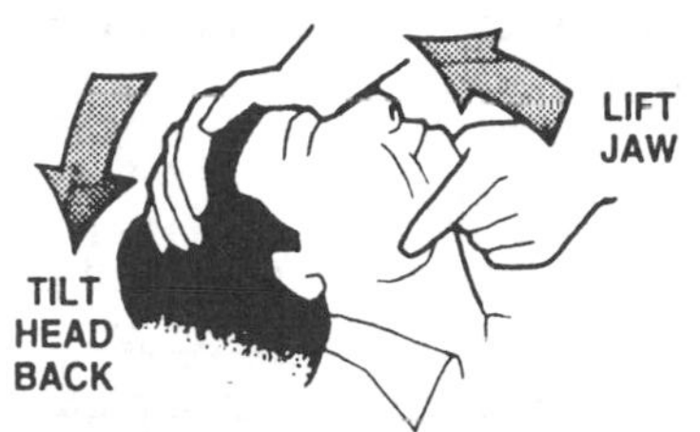

BREATHING MAY RESTART ... IF NOT ...

3 APPLY RESCUE BREATHING

START WITH FOUR QUICK DEEP BREATHS

SEAL NOSE AND BLOW INTO MOUTH

or

SEAL MOUTH AND BLOW INTO NOSE

KEEP FINGERS ON JAW BUT CLEAR OF THROAT

MAINTAIN HEAD POSITION

AFTER BLOWING INTO MOUTH or NOSE, WATCH CASUALTY'S CHEST FALL AS YOU BREATHE IN

REPEAT EVERY 5 SECS

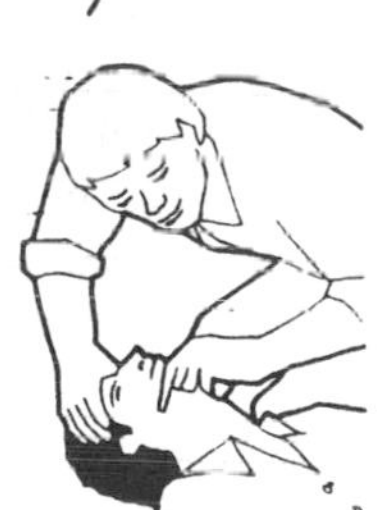

AFTER FIRST FOUR BREATHS TEST FOR RECOVERY SIGNS

1. PULSE PRESENT?
2. PUPILS LESS LARGE?
3. COLOUR IMPROVED?

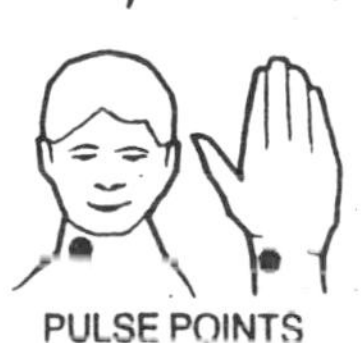

4 IF NONE, COMBINE RESCUE BREATHING & HEART COMPRESSION

PLACE CASUALTY ON A FIRM SURFACE

COMMENCE HEART COMPRESSION

HEEL OF HAND ONLY ON LOWER HALF OF BREASTBONE
OTHER HAND ON TOP, FINGERS OFF CHEST

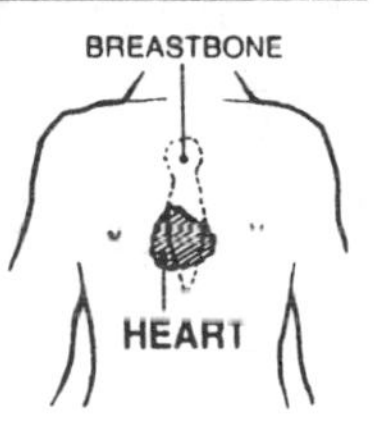

KEEP ARMS STRAIGHT AND ROCK FORWARD TO DEPRESS CHEST 1½ INCHES (4 cm)

APPLY 15 COMPRESSIONS ONE PER SECOND ... then GIVE TWO BREATHS

RE-CHECK PULSE ...
IF STILL ABSENT CONTINUE WITH 15 COMPRESSIONS TO TWO BREATHS

IF PULSE RETURNS CEASE COMPRESSIONS BUT CONTINUE RESCUE BREATHING

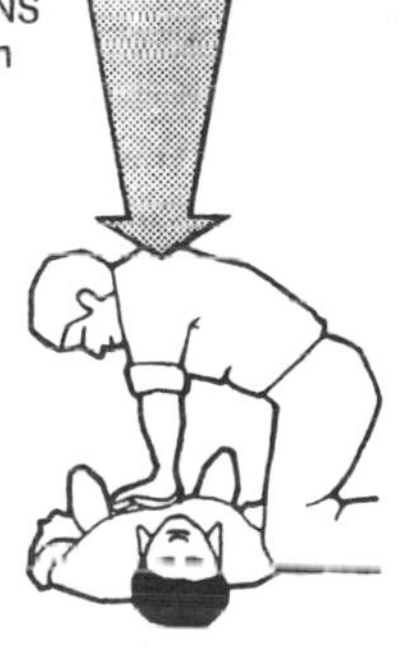

• **FIG 3.4 Resuscitation procedure**
Source: RoSPA

DISPLAY SCREEN EQUIPMENT

Legal provisions relating to display screen equipment are covered by the Health and Safety (Display Screen Equipment) Regulations 1992. (See further Chapter 9.) 'Display screen equipment' is defined in the regulations as meaning 'any alphanumeric or graphic display screen, regardless of the display process involved'.

In the last decade considerable attention has been given to the problem of operator stress and ill health associated with the use of display screen equipment, in particular that group of disorders known as 'work-related upper limb disorders'.

Effects on health

The more general effects on health include visual fatigue, general fatigue and operational stress. Visual fatigue or eye strain is related to glare from the display and the continual need to focus and refocus from screen to source material and back again.

General fatigue can be brought about by poor environmental control, temperature variations, the sheer monotony of some display screen tasks, inefficient machine response, work pressure, poor ergonomic design of controls and displays, and screen flicker.

Operational stress may take many forms, and can include backache, neck and shoulder pains associated with poor chair and workstation design in relation to controls and displays, insufficient leg room and the need to adjust body position regularly, noise from the unit and ancillary equipment, excessive heat and inadequate ventilation. The degree of operator stress will vary according to age, sex, physical build, attitude to the task, current level of visual acuity and general health.

Work-related upper limb disorders and RSI are further associated with the use of display screen equipment. (See earlier in this chapter.)

There is no scientific or medical evidence to support the views that work at a display screen can result in epilepsy or facial dermatitis, or that there may be risks to pregnant women due to electromagnetic radiation.

Control strategies

The Schedule and Annexes to the regulations provide sound guidance on the control strategies necessary to avoid operator stress. The Schedule is reproduced below.

The Schedule

(Which sets out the minimum requirements for workstations which are contained in the Annex to Council Directive 90/270/EEC on the minimum safety and health requirements for work with display screen equipment)

Extent to which employers must ensure that workstations meet the requirements laid down in this Schedule

1. An employer shall ensure that a workstation meets the requirements laid down in the Schedule to the extent that:
   (a) those requirements relate to a component which is present in the workstation concerned;
   (b) those requirements have effect with a view to securing the health, safety and welfare of persons at work; and
   (c) the inherent characteristics of a given task make compliance with those requirements appropriate as respects the workstation.

Equipment

2. (a) **General comment**
   The use as such of the equipment must not be a source of risk for operators or users.

   (b) **Display screen**
   The characters on the screen shall be well-defined and clearly formed, of adequate size and with adequate spacing between the characters and lines.

   The image on the screen should be stable, with no flickering or other forms of instability.

   The brightness and the contrast between the characters and the background shall be easily adjustable by the operator or user, and also be easily adjustable to ambient conditions.

   The screen must swivel and tilt easily and freely to suit the needs of the operator or user.

   It shall be possible to use a separate base for the screen or an adjustable table.

   The screen shall be free of reflective glare and reflections liable to cause discomfort to the operator or user.

   (c) **Keyboard**
   The keyboard shall be tiltable and separate from the screen so as to allow the operator or user to find a comfortable working position avoiding fatigue in the hands or arms.

   The space in front of the keyboard shall be sufficient to provide support for the hands and arms of the operator or user.

   The keyboard shall have a matt surface to avoid reflective glare.

   The arrangement of the keyboard and the characteristics of the keys shall be such as to facilitate the use of the keyboard.

   The symbols on the keys shall be adequately contrasted and legible from the design working position.

   (d) **Work desk or work surface**
   The work desk or work surface shall have sufficiently large, low-reflectance surface and allow a flexible arrangement of the screen, keyboard, documents and related equipment.

   The document holder shall be stable and adjustable and shall be positioned so as to minimise the need for uncomfortable head and eye movements.

There shall be adequate space for operators or users to find a comfortable position.

(e) **Work chair**
The work chair shall be stable and allow the operator or user easy freedom of movement and a comfortable position.

The seat shall be adjustable in height.

The seat back shall be adjustable in both height and tilt.

A footrest shall be made available to any operator or user who wishes one.

Environment

3. (a) **Space requirements**
The workstation shall be dimensioned and designed so as to provide sufficient space for the operator or user to change position and vary movements.

(b) **Lighting**
Any room lighting or task lighting provided shall ensure satisfactory lighting conditions and an appropriate contrast between the screen and the background environment, taking into account the type of work and the vision requirements of the operator or user.

Possible disturbing glare and reflections on the screen or other equipment shall be prevented by co-ordinating workplace and workstation layout with the positioning and technical characteristics of the artificial light source.

(c) **Reflections and glare**
Workstations shall be so designed that sources of light, such as windows and other openings, transparent or translucid walls, and brightly coloured fixtures or walls cause no direct glare and no distracting reflections on the screen.

Windows shall be fitted with a suitable system of adjustable covering to attenuate the daylight that falls on the workstation.

(d) **Noise**
Noise emitted by equipment belonging to any workstation shall be taken into account when a workstation is being equipped, with a view in particular to ensuring that attention is not distracted and speech is not disturbed.

(e) **Heat**
Equipment belonging to any workstation shall not produce excess heat which could cause discomfort to operators or users.

(f) **Radiation**
All radiation with the exception of the visible part of the electromagnetic spectrum shall be reduced to negligible levels from the point of view of the protection of operators' or users' health and safety.

(g) **Humidity**
An adequate level of humidity shall be established and maintained.

Interface between computer and operator/user

4. In designing, selecting, commissioning and modifying software, and in designing tasks using display screen equipment, the employer shall take into account the following principles:

   (a) software must be suitable for the task;

   (b) software must be easy to use and, where appropriate, adaptable to the level of knowledge or experience of the operator or user; no quantitative or qualitative checking facility may be used without the knowledge of the operators or users;

   (c) systems must provide feedback to operators or users on the performance of those systems;

   (d) systems must display information in a format and at a pace which are adapted to operators and users;

   (e) the principles of software ergonomics must be applied, in particular to human data processing.

It should be appreciated that the comprehensive requirements of the above Schedule to the regulations are mainly directed at reducing operator stress. General measures to prevent and/or avoid RSI are dealt with earlier in this chapter. Personal measures to avoid the onset of RSI include:

- taking regular breaks away from the keyboard and screen
- planning to avoid sudden surges of work
- ensuring the appropriate equipment or furniture is used
- keeping the flow of work to manageable proportions
- making sure jobs have some form of variety in order to eliminate repetitive movements.

Biological health hazards

A number of occupational diseases are transmissible from animal to Man. This group of diseases is known as the **zoonoses**, and may include diseases contracted through viral infection. Other diseases of a biologically-induced nature can be transmitted by various bacilli, leptospira and fungi.

These diseases and their causative organisms are classified below in Table 4.1.

TABLE 4.1 Occupational diseases of a biologically-induced nature

Disease	*Causative organism*
1 Zoonoses	
Anthrax	Bacillus anthracis
Glanders	Bacillus mallei or Pfeifferella mallei
Brucellosis	Brucella abortus
Q fever	Ricketsia burneti
Orf	Specific virus
Psittacosis	Bedsonia virus
2 Bacilli	
Legionnaire's disease	Legionella pneumophila
Pontiac fever	Legionella pneumophila
3 Leptospira	
Weil's disease	Leptospira icterohaemorrhagica
4 Fungi	
Aspergillosis	Micropolyspora faeni
Bagassosis	Aspergillus fumigatus

Whilst the incidence of such diseases is low, there is always some degree of risk to anyone working with animals, especially veterinary surgeons, meat inspectors, slaughtermen, pet shop workers, people working in zoos, artificial inseminators and farmers. Anthrax, glanders, Weil's disease (leptospirosis) and brucellosis are prescribed occupational diseases.

THE ZOONOSES

Anthrax

This is a disease which may occur in Man and certain animals e.g. cattle and sheep, as a result of infection by **Bacillus anthracis**, a spore-forming organism which, although killed by boiling for 10 minutes, may survive for years in the soil and in animal remains. Cattle are the main source of infection, and infection in Man may occur through contact with fresh infective material containing the bacillus. People at risk include agricultural workers, veterinary surgeons, knackers, slaughtermen, and those working with dried animal products such as hides, skins, hair, wool, hooves, bonemeal and contaminated implements.

Infection in Man may be of the cutaneous type (malignant pustule) or internal e.g. the pulmonary form (wool sorter's disease). The disease manifests itself in almost every case as a grave toxaemia, with headache, shivering, muscle and joint pains, nausea, vomiting and collapse, together with additional symptoms depending on the site and type of infection.

Malignant pustule is the more common form of anthrax. Infection takes place through cuts and abrasions on the skin. After an incubation period of one–four days an irritant pimple develops. This rapidly enlarges and breaks down with a black necrotic centre. The lesion may be ringed with small vesicles and inflammatory swelling. Local lymph nodes may be slightly enlarged. In 90 per cent of cases the pustule is situated on some exposed part of the body such as the face or neck, and in such cases the intense oedema may be fatal.

Internal anthrax (wool sorter's disease) takes place through ingestion or inhalation of the bacilli. In these cases, even more than in external cases, the general intense toxaemia, with sudden vertigo, somnolence, dyspnoea (difficulty in breathing), croup and marked prostration, is prevalent, and death may ensue. In typical pulmonary cases, there is widespread congestion and oedema or an atypical pneumonia with blood-stained frothy sputum. If untreated, death occurs from septicaemia in the first few days.

The majority of cases of anthrax are treated successfully with penicillin.

Glanders fever

Glanders fever or 'farcy' is a disease of horses, mules and donkeys. The infecting organism is **Bacillus mallei** or **Pfeifferella mallei**. Infection in Man is now rare, but is always caused by contact with infected material. The disease occurs in both acute and chronic form.

In the acute form there is an incubation period of two to three days before general malaise is experienced, together with headaches, anorexia and joint pains. The site of infection becomes ulcerated and there is marked lymphangitis (inflammation of the lymph vessels). Nodular abscesses form along the lymphatic vessels and these break down to form painful ulcers. There is a marked high fever, highest between the sixth and twelfth days, after which time eruptions appear on the face and on the nasal, palatal and pharyngeal mucosae. The lesions typically begin as patches which eventually enlarge and form pustules. The pustules ulcerate with the destruction of bone and cartilage or produce a thick blood-stained discharge. A form of arthritis may also occur with the development of abscesses in the muscles.

The chronic form is rarer than the acute form, but is again characterised by the formation of abscesses, which break down to form painful ulcers. The lungs may be involved in terms of pneumonia, pleural effusion, lung abscesses and empyema (a collection of pus in a natural body cavity e.g. in the space between the lung and outer wall of the chest). The disease runs a long course and an acute phase may supervene at any time.

The disease is successfully treated through a course of antibiotics.

Brucellosis

Also known as 'undulant fever', brucellosis in Man is caused by contact with infected animals. Three species of the organism account for most human disease. These species show an affinity for a particular animal host, so **that Brucella abortus** is found in cattle, **Brucella melitensis** in sheep and goats, and **Brucella suis** in pigs. The disease may be contracted by persons working in slaughterhouses or among those handling infected meat products or the by-products and waste from slaughtering. Veterinary surgeons and meat inspectors are an outstanding high-risk group.

The routes of infection can be through inhalation, ingestion and direct contact with infected material e.g. the uterus of an infected animal, direct contact being the most important. In the latter case, this occurs usually through handling the placenta or foetal parts during the delivery of a calf or in post-mortem examinations. The organism gains access through cuts and abrasions in the skin or through the mucous membranes, including the conjunctivae.

Brucellosis takes two forms, the acute attack and the chronic condition. In acute cases, onset may be gradual with non-specific signs such as headache, joint pains, fever, insomnia and low back pain, or it may be abrupt with fever, rigors and prostration. Usually the disease subsides within two weeks and the patient makes a complete recovery. Some patients will continue, however, to have intermittent bouts of fever, back pain, a feeling of lethargy and depression, which may last for several months or years.

Chronic brucellosis has all the symptoms of an acute attack i.e. lassitude, malaise, joint pains and prolonged depression. There is not always a history of an acute attack and, in many cases, the occupation of the patient may be the only clue in diagnosis e.g. as a stockman on a farm. In chronic brucellosis there may be complications including endocarditis (inflammation of the heart lining) and spondylitis (inflammation of the vertebrae).

Most cases of brucellosis respond to treatment with antibiotics.

Q fever (Query fever)

This is an infection caused by an organism, **Rickettsia burneti** or **Coxiella burneti**. The infection is found most frequently in farm workers who contract the disease from sheep and cows by the inhalation of infected dust or by drinking infected raw milk. Veterinary surgeons, meat inspectors and slaughtermen in abattoirs are particularly high risk groups in this case.

The symptoms of infection are very similar to those of influenza and it is common for cases of Q fever to be diagnosed as such. Typically, the illness begins with fever accompanied by shivering, sweating and backache, inflammation of the throat and suffused conjunctivae. In many cases, the patient has an unproductive cough, photophobia and muscle pains.

Like other rickettsial diseases, Q fever responds well to treatment with antibiotics.

Orf

Known also as 'contagious pustular dermatitis', orf is a viral infection of sheep which is transmitted occasionally to abattoir workers and animal handlers. The disease takes the form of a mild skin rash occurring at the site of infection.

Clinical signs appear 4–12 days after infection, with the development of a red macule (a spot level with the surface of the surrounding skin) or papule (a raised spot on the surface of the skin). This enlarges until it becomes 1–4cm in diameter containing first clear fluid and then pus. There may be some local tenderness and lymphadenitis, and the lesion is sometimes painful. Healing is usually complete within four–six weeks.

Good standards of personal hygiene, together with the prompt treatment and covering of skin lesions, are important factors in preventing infection.

Psittacosis

This is a pneumonia-like condition caused by infection with the **Bedsonia** virus carried by game, poultry and other birds, such as parrots. Psittacosis can be fatal in Man if untreated.

The illness has a sudden effect after an incubation period of two–three weeks and is characterised by initial fever, headache and lethargy. These symptoms are followed by pulmonary symptoms several days later, including non-productive cough and shallow breathing. Elderly people may die as a result of the infection.

BACILLARY INFECTIONS

Legionnaire's disease and Pontiac fever

Legionnaire's disease is so called because of its association with the Annual Convention of the Pennsylvanian Department of the American Legion held in Philadelphia in 1976. This Convention resulted in reports of 12 deaths through the inhalation of aerosols containing a specific bacterium, **Legionella pneumophila**, resulting in pneumonia-like symptoms among the people infected.

A similar situation arose two years later in the County Health Department at Pontiac, Michigan. Here it was established that the water-cooled air-conditioning system was designed in such a way that waste water and heat were discharged at roof level at a point adjacent to the air inlet to the system. Investigations showed that Pontiac fever and Legionnaire's disease were caused by the same bacterium.

Legionella bacteria are widely distributed in the environment and occur in at least ten different forms. They are commonly encountered in water cooling systems, rivers, streams, ponds, lakes and in the soil. Most reported cases occur in the 40 to 70 years age group.

Initial symptoms of the disease include high fever, chills, headache and muscle pain, and there is an incubation period which may range from two to ten days, but usually three to six days. After a short period, a dry cough develops and most patients suffer difficulty with breathing. About one-third of patients also develop diarrhoea or vomiting and about 50 per cent of patients may become confused or delirious. The disease may not always be severe and mild cases may be recognised which would probably have escaped diagnosis except for the increased awareness of this disease among doctors and managers.

Conditions that affect the proliferation of **Legionella** bacteria include:

- the presence of sludge, scale, rust, algae and organic particulates which, although the ecology of **Legionella** is not fully understood, are thought to provide nutrients for growth
- water temperatures in the range 20°C to 45°C which favours growth.

Legionella is frequently found in many recirculating and hot water systems, particularly large complex systems, such as those incorporated in large multi-storey office blocks, factories and hospitals. Particular sites for bacterial growth are air-conditioning systems, cooling towers, water standing in ductwork and condensate trays, humidifiers, hot and cold water storage tanks, calorifiers, pipework and plant.

Regular sampling of water in these types of installation is recommended together with frequent disinfection of the system. Further guidance is available in HSE Guidance Note EH 48 'Legionnaire's Disease' and publications by local water authorities and environmental health departments.

LEPTOSPIRAL INFECTIONS

Leptospirosis

This disease is also known as 'leptospiral jaundice', 'spirochaetal jaundice', 'spirochaetosis icterohaemorrhagica', 'mud fever' and 'Weil's disease'. It is a feverish condition caused by the organism **Leptospira icterohaemorrhagica**, commonly found in rats, which are the source of human infection. Infection may be due to ingestion of food or water contaminated with the urine of infected rats; alternatively, it may enter via the skin or through the mucous membranes of the eyes, nose and mouth. The disease sometimes occurs among workers in rat-infested locations, such as mines, sewers, canals, slaughterhouses and fish docks, where they may be brought into contact with infected water, mud and slime. A less virulent form of the disease is caused by **Leptospira canicola**, a canine leptospirosis.

4

After an incubation period of 6–12 days there is an abrupt onset of high fever, rigors, headache, muscular pain and vomiting, accompanied by prostration. At this time, the leptospires multiply in the blood and may be carried to and affect any organ. Conjunctival haemorrhages are common together with a body rash, often accompanied by petechial (pinpoint-sized) haemorrhages in the skin. There may be mild liver damage and jaundice is common two–five days after the onset of fever. There is usually a steady improvement after the second week of the illness and mild cases recover completely without specific treatment. Fatalities are rare.

Prevention and control is based on effective rodent control and improved structural hygiene and cleaning in infested areas. Operators should maintain high standards of personal hygiene and be provided with regular changes of overalls in particular. In addition, they should be instructed in the dangers of contracting leptospirosis, the symptoms, the personal precautions necessary.

FUNGAL INFECTIONS

Aspergillosis

Aspergillosis is associated with exposure to mouldy hay or other mouldy vegetable produce which can result in pulmonary disease. It is an all-embracing term describing types of extrinsic allergic alveolitis, an asthma-like condition, caused by inhalation of the spores of the **Aspergillus** fungus, principally **Aspergillus fumigatus** and **Aspergillus niger**. All of these spores are encountered as mould on fibres such as jute, flax, hemp, straw and hay.

Farmer's lung

Mouldy straw and hay encourages the growth of certain moulds, in particular **Micropolyspora faeni**. When handled in the field or in a barn, clouds of dust containing these spores are liberated into the surrounding air, and are subsequently inhaled by workers. However, not all workers develop farmer's lung as the disease is the result of individual hypersensitivity due to an antigen present in the dust of mouldy hay and other vegetable matter.

It is characterised, along with many other similar conditions, such as mushroom picker's lung and malt worker's lung, by an influenza-like illness, during which the person feels generally unwell, has pain in the limbs and is feverish. The patient will also have a dry cough and dyspnoea.

Farmer's lung is one form of extrinsic allergic alveolitis, an inflammatory condition of the lung tissue associated with hypersensitivity to the spores of mouldy hay. It is usually a transitory condition where the symptoms abate after three–four days. It is a prescribed occupational disease.

Bagassosis

The term **bagasse** is of French origin and describes the fibres of cane sugar following extraction of the sugar. These waste fibres are used in a number of processes, such as the manufacture of various forms of fibreboard.

The fibres contain many different species of fungal spores including **Aspergillus fumigatus** which, if inhaled, produce a pulmonary condition similar to farmer's lung.

OTHER INFECTIONS

Viral hepatitis

Hepatitis, inflammation of the liver, is associated with a number of infections. The disease takes two forms, namely epidemic jaundice and serum hepatitis.

Epidemic jaundice (Hepatitis A) is generally spread from person to person by direct contact or can be contracted from contaminated food or water supplies.

Serum hepatitis (Hepatitis B), on the other hand, occurs commonly among members of the medical profession and allied professions, the risk being greatest among those who handle blood and blood products and who work in renal dialysis units. Other groups who are at risk include hospital porters and refuse collectors, who may come into contact with contaminated material, particularly from carelessly discarded syringes in polythene refuse sacks.

The symptoms of the disease include malaise, myalgia (muscle pain), headache, nausea, vomiting, anorexia, abdominal pain and pruritis (itching). The patient becomes jaundiced and the liver is enlarged. Generally, the disease runs a mild course although, in a small number of cases, chronic infectious hepatitis may follow, leading to cirrhosis and possibly death.

Control among those at risk is through injections of gammaglobulin. High standards of personal hygiene are also required, including the use of disposable gloves, among those who may be treating patients and coming into contact with infected blood and materials.

AIDS

AIDS (acquired immune deficiency syndrome) is an important public health hazard that has attracted widespread publicity. It is a disease caused by a specific virus, HIV (human immunodeficiency virus).

A person is said to be an AIDS **sufferer** if, following infection by the AIDS virus, a particular form of cancer, Kaposi's sarcoma, or serious infection, a form of pneumonia caused by the organism, **Pneumocystis carinii**, has developed and which is the result of the breakdown of the body's normal defences.

AIDS **carriers**, on the other hand, are people who are infected with the virus but who may show no symptoms and, as a result, may be unaware of their infection.

People can become infected only by intimate physical contact or contact with body fluids, as in sexual intercourse, or by the use of contaminated syringes and needles for drug injections. It may also be transmitted from an infected mother to her unborn child.

People infected with AIDS may demonstrate a number of related symptoms, such as swollen glands, extreme fatigue, recurrent fever, rapid weight loss, persistent diarrhoea, white spots (thrush) in the mouth and unexplained bruising and/or bleeding.

AIDS and first aiders

The AIDS virus and other infections like hepatitis B are carried in the bloodstream of infected persons but are not easily transmitted to others unless the blood is injected or where large quantities of infected blood come into contact with broken skin. However, there is no conclusive evidence that infection has occurred as a result of infected blood coming into contact with a person's skin, for example, of the hands, as could occur when carrying out first aid treatment. Nevertheless, any blood that splashes on unprotected skin should be washed away with soap and hot water as soon as possible.

The AIDS virus has been found only occasionally in saliva and in very small quantities when compared with blood. No AIDS infection is known to have occurred as a result of undertaking mouth-to-mouth resuscitation. The risk to a first aider of suffering any infection when giving mouth-to-mouth resuscitation is extremely small and should not discourage a prompt response in a life-saving emergency situation.

The use of rigid airways by those unskilled in their use may cause bleeding, increasing rather than diminishing any risk of infection.

The chief medical advisers of the voluntary aid societies i.e. Saint John Ambulance, Saint Andrew's Ambulance Association and the British Red Cross Society, after consultation with the Chief Medical Officer of the Department of Health, therefore see no grounds for recommending changes in resuscitation techniques, or procedures for arresting bleeding, as described in the First Aid Manual, because of AIDS or the virus associated with it.

The key features of AIDS

1. AIDS means acquired immune deficiency syndrome.
2. It is a viral infection – human immunodeficiency virus (HIV).
3. Transmission is principally through blood and semen, but the virus has been isolated in saliva and tears.
4. Infection only occurs when the virus enters the bloodstream.
5. The virus destroys the body's natural immune system leaving it vulnerable to attack by other organisms.
6. Infected people may be carriers, possibly for life, and are classed as being 'HIV positive'.
7. Infected people may develop the condition known as 'full-blown AIDS'.
8. Speed of onset and development are affected by general health, lifestyle, diet and genetic make-up.

The AIDS test

The Health Education Authority leaflet 'The AIDS Test' provides guidance for anyone who is thinking of having a blood test known as the 'HIV antibody test'.

Toxicology, epidemiology and pathology

In the study of the principles of toxicology, epidemiology and pathology, the following terms are significant:

- **Toxicology**
  - — the study of the body's responses to toxic substances.
- **Toxicity**
  - — the ability of a chemical molecule to produce injury once it reaches a susceptible site in or on the body
  - — the quantitative study of the body's responses to toxic substances.
- **Intoxication**
  - — the general state of harm caused by the effects of a toxic substance.
- **Detoxification**
  - — the process in the body when decomposition of toxic substances occurs to produce harmless substances which are eliminated from the body.
- **Epidemiology**
  - — the study of the distribution of diseases in different groups of people.
- **Pathology**
  - — the science of diseases
  - — the natural history of disease
  - — the study of abnormal changes in the body and their causes.

THE EFFECTS OF EXPOSURE TO TOXIC SUBSTANCES

The effects on the body of exposure to toxic substances vary considerably. These various effects are outlined below.

1 *Acute effect* a rapidly produced effect following a single exposure to an offending agent.

2 *Chronic effect* an effect produced as a result of prolonged exposure or repeated exposures of long duration. Concentrations of the offending agent may be low in both cases.
3 *Sub-acute effect* a reduced form of acute effect.
4 *Progressive chronic effect* an effect which continues to develop after exposure ceases.
5 *Local effect* an effect usually confined to the initial point of contact. The site may be the skin, mucous membranes of the eyes, nose or throat, liver, bladder, etc.
6 *Systemic effect* such effects occur in parts of the body other than at the initial point of contact, and are associated with a particular body system e.g. respiratory system, central nervous system.

TOXIC SUBSTANCES – ROUTES OF ENTRY

Inhalation

Inhalation of toxic substances, in the form of a dust, gas, mist, fog, fumes or vapour, accounts for approximately 90 per cent of all ill health associated with toxic substances. The results may be **acute** (immediate) as in the case of gassing accidents e.g. chlorine, carbon monoxide, or **chronic** (prolonged, cumulative) as in the case of exposure to chlorinated hydrocarbons, lead compounds, benzene, numerous dusts, which produce pneumoconiosis, mists and fogs, such as those from paint spray, oil mist, and fumes, such as those from welding operations.

Pervasion

The skin, if intact, is proof against most, but not all, inputs. There are certain substances and micro-organisms which are capable of passing straight through the intact skin into underlying tissue, or even into the bloodstream, without apparently causing any changes in the skin.

The resistance of the skin to external irritants varies with age, sex, race, colour and, to a certain extent, diet. Pervasion, as a route of entry, is normally associated with occupational dermatitis, the causes of which may be broadly divided into two groups:

- **Primary irritants** are substances which will cause dermatitis at the site of contact if permitted to act for a sufficient length of time and in sufficient concentrations e.g. strong alkalis, acids and solvents.
- **Secondary cutaneous sensitisers** are substances which do not necessarily cause skin changes on first contact, but produce a specific sensitisation of the skin. If further contact occurs after an interval of, say, seven days or more, dermatitis will develop at the site of the second contact. Typical skin sensitisers are plants, rubber, nickel and many chemicals.

It should be noted that, for certain people, dermatitis may be a manifestation of psychological stress, having no relationship with exposure to toxic substances (an endogenous response).

Ingestion

Certain substances are carried into the gut from which some will pass into the body by absorption. Like the lung, the gut behaves as a selective filter which keeps out many, but not all, harmful agents presented to it.

Injection/implantation

A forceful breach of the skin, frequently as a cause of injury, can carry substances through the skin barrier.

Toxic substances are widely used in industry. Some indications of the most common hazards and the occupations associated with them are listed in the Social Security (Industrial Injuries) (Prescribed Diseases) Regulations 1985.

DOSE–RESPONSE RELATIONSHIP

A basic principle of occupational disease prevention rests upon the reality of threshold levels of exposure for the various hazardous agents, below which, Man can cope successfully without significant threat to his health. This concept derives from the quantitative characteristic of the dose-response relationship, according to which there is a systematic downward change in the magnitude of Man's response as the dose of the offending agent is reduced.

Dose = Level of environmental contamination x Duration of exposure

With many dusts, for instance, the body's response is directly proportional to the dose received over a period of time – the greater the dose, the more serious the condition, and vice versa. However, in the case of airborne contaminants, such as gases or mists, there is a concentration in air or dose below which most people can cope reasonably well. Once this concentration in air is reached (threshold dose), some form of body response will result. This concept is most important in the correct use and interpretation of occupational exposure limits (formerly known as 'threshold limit values').

In Fig 5.1, Diagram A shows a typical direct dose-response relationship, which is a feature of exposure to many dusts. In Diagram B the dose-response curve reaches a level of 'no response' at a point greater than zero on the dose axis. This point of cut-off identifies the threshold dose which was the original basis for the setting of threshold limit values (occupational exposure limits).

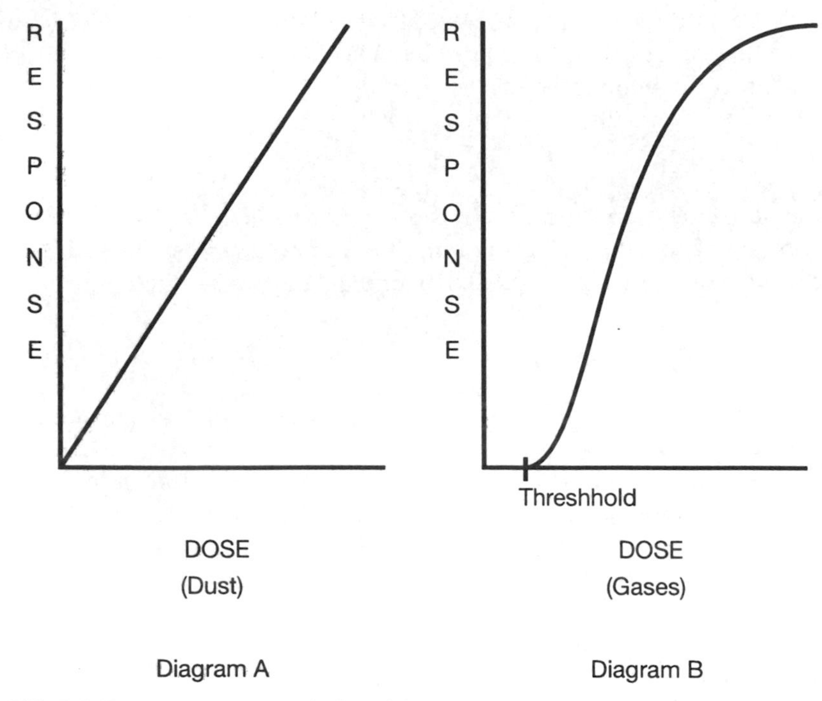

• **FIG 5.1 Dose-response relationship**

TARGET ORGANS AND TARGET SYSTEMS

Certain substances have a direct or indirect effect on specific body organs (target organs) and body systems (target systems). Typical examples are:

- *target organs* lungs, liver, brain, skin, bladder
- *target systems* central nervous system, circulatory system, reproductive system.

PROTECTIVE MECHANISMS – PARTICULATE MATTER

The protective mechanisms within the body respond largely according to the shape and size of particulate matter which may be inhaled. Such mechanisms are as follows (see also Fig 5.2).

Nose

The coarse hairs in the nose, assisted by mucus from the nasal lining, act as a filter, trapping the larger particles of dust.

Ciliary escalator

The respiratory tract consists of the trachea (windpipe) and bronchi which branch to the lungs. The lining of the trachea consists of quite tall cells, each of which has a cilium growing from its head (ciliated epithelium). These cilia exhibit a wave-like motion so that a particle falling onto the cilia is returned back to the throat. Mucus helps these particles to stick.

Macrophages/phagocytes

These are wandering scavenger cells. They have an irregular outline and large nucleus, and can move freely through body tissue, engulfing bacteria and dust particles. They secrete hydrolytic enzymes which attack the foreign body.

Lymphatic system

This is a drainage system which acts as a clearance channel for the removal of foreign bodies, many of which are retained in the lymph nodes throughout the body. In certain cases, a localised inflammation will be set up in the lymph node.

Tissue response

A typical example of tissue response is in byssinosis ('Monday fever'), a chest condition of cotton workers, where the lung becomes sensitised to cotton dust through continuing exposure.

THE PHYSICAL STATE OF HAZARDOUS SUBSTANCES

In any evaluation of the risks associated with hazardous substances, it is necessary to consider the actual form taken by that substance. Substances carried in air are **aerosols**.

Forms of aerosol

Features of the various forms of aerosol are outlined below.

Dusts

Dust is an aerosol composed of solid inanimate particles (International Labour Organisation). Dusts are solid airborne particles, often created by operations such as grinding, crushing, milling, sanding and demolition. Two of the principal dusts encountered in industry are asbestos and silica.

Dusts may be:

- *fibrogenic* they cause fibrotic changes to lung tissue e.g. silica, cement dust, coal dust and certain metals
- *toxic* they eventually poison the body systems e.g. arsenic, mercury, beryllium, phosphorus and lead.

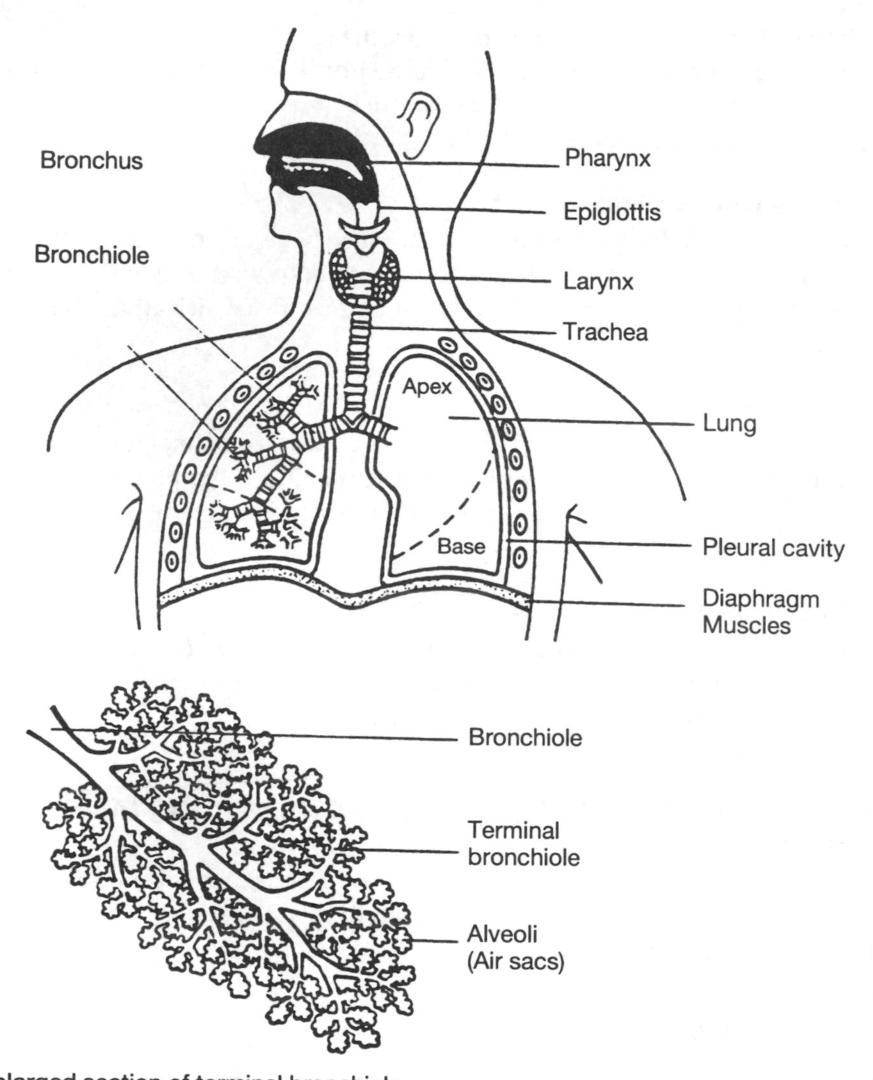

• FIG 5.2 The human respiratory system – protective mechanisms against inhalation of solid materials

Mists

A mist comprises airborne liquid droplets, a finely dispersed liquid suspended in air. Mists are mainly created by spraying, foaming, pickling and electroplating. Danger arises most frequently from acid mist produced in industrial treatment processes e.g. oil mist, chromic acid mist.

Fumes

These are fine solid particulates formed from the gaseous state usually by vaporisation or oxidation of metals e.g. lead fumes. Fumes usually form an oxide when in contact with air. They are created by industrial processes which involve the heating and melting of metals, such as welding, smelting and arc air gouging. A common fume danger is lead poisoning associated with the inhalation of lead fumes.

Gases

These are formless fluids usually produced by chemical processes involving combustion or by the interaction of chemical substances. A gas will normally seek to completely fill the space into which it is liberated. A classic gas encountered in industry is carbon monoxide. Certain gases such as acetylene, hydrogen and methane are particularly flammable.

Vapours

A vapour is the gaseous form of a material normally encountered in a solid or liquid state at normal room temperature and pressure. Typical examples are solvents, such as trichloroethylene, which release vapours when the container is opened. Other liquids produce a vapour on heating, the amount of vapour being directly related to the boiling point of that particular liquid.

A vapour contains very minute droplets of the liquid. However, in the case of a **fog**, the liquid droplets are much larger.

Smoke

Smoke is a product of incomplete combustion, mainly of organic materials. It may include fine particles of carbon in the form of ash, soot and grit that are visibly suspended in air.

OCCUPATIONAL EXPOSURE LIMITS

Threshold limit values (TLVs) are values published by the American Conference of Government Industrial Hygienists and adopted for use in the UK by the HSE. They refer to airborne concentrations of substances and represent conditions under which it is believed that nearly all workers may be repeatedly exposed, day after day, without adverse effect. TLVs refer to time-weighted average concentrations for a 7- or 8-hour work day and a 40-hour working week. They have now been replaced by occupational exposure limits (OELs).

HSE Guidance Note EH40 'Occupational Exposure Limits' gives details of OELs which should be used for determining the adequacy of control of exposure by inhalation of substances hazardous to health. These limits form part of the requirements of the Control of Substances Hazardous to Health (COSHH) Regulations 1994.

Under regulation 2 of the COSHH Regulations (as amended) a **substance hazardous to health** is defined as:

'any substance (including any preparation) which is:

(a) a substance which is listed in the Approved Supply List, and classified on the basis of health effects, as dangerous for supply within the meaning of the Chemicals (Hazard Information and Packaging for supply) Regulations 1994 and for which the general indication of nature of risk is specified as very toxic, toxic, harmful, corrosive or irritant;

(b) a substance for which a maximum exposure limit (MEL) is specified in Schedule 1 or for which the HSC has approved an occupational exposure standard (OES);

(c) a biological agent;

(d) dust of any kind, when present at a substantial concentration in air; and

(e) a substance, not being a substance mentioned in sub-paragraphs (a) to (d) above, which creates a hazard to the health of any person which is comparable with the hazards created by substances mentioned in those sub-paragraphs'.

The advice given in Guidance Note EH40 should be taken in the context of the requirements of the COSHH Regulations, especially regulation 6 (health risk assessments), regulation 7 (control of exposure), regulations 8 and 9 (use and maintenance of control measures) and regulation 10 (monitoring of exposure). Additional guidance may be found in the COSHH General Approved Code of Practice.

Legal requirements

Regulation 2(1) of COSHH defines the following terms:

'the **maximum exposure limit**' for a substance hazardous to health means the maximum exposure limit for that substance set out in Schedule 1 in relation to the reference period specified therein when calculated by a method approved by the HSC.

'the **occupational exposure standard**' for a substance hazardous to health means the standard approved by the HSC for that substance in relation to a specified reference period when calculated by a method approved by the HSC.

Regulation 7(4) of COSHH requires that where there is exposure to a substance for which an MEL is specified in Schedule 1, the control of exposure, so far as inhalation of that substance is concerned, shall only be treated as being adequate if the level of exposure is reduced so far as is reasonably practicable and in any case below the MEL.

Regulation 7(5) of COSHH requires that, without prejudice to the generality of regulation 7(1), where there is exposure to a substance for which an OES has been approved, the control of exposure shall, so far as inhalation of that substance is concerned, be treated as being adequate if:

- the OES is not exceeded, or
- where the OES is exceeded, the employer identifies the reasons for the standard being exceeded and takes appropriate action to remedy the situation as soon as is reasonably practicable.

Units of measurement

The lists of OELs given in the Guidance Note, unless otherwise stated, relate to personal exposure to substances hazardous to health in the air of the workplace. Concentrations of gases and vapours in air are usually expressed as parts per million (ppm), a measure of concentration by **volume**, as well as in milligrams per cubic metre of air (mg/m^3), a measure of concentration by **mass**. Concentrations of airborne particles (fumes, dust, etc.) are usually expressed in milligrams per cubic metre, with the exception of mineral fibres, which are expressed as fibres per millilitre of air.

MAXIMUM EXPOSURE LIMITS AND OCCUPATIONAL EXPOSURE STANDARDS

5

Maximum exposure limits (MELs)

MELs are listed in both Schedule 1 of the COSHH Regulations and Table 1 of Guidance Note EH40.

An MEL is the maximum concentration of an airborne substance, averaged over a reference period, to which employees may be exposed by inhalation under any circumstances and is specified, together with the appropriate reference period, in Schedule 1 of the COSHH Regulations.

Regulation 7(4) of COSHH, when read in conjunction with regulation 16, imposes a duty on the employer to take **all reasonable precautions** and to exercise **all due diligence** to achieve these requirements.

In the case of substances with an 8-hour long-term reference period, unless the assessment carried out in accordance with regulation 6 shows that the level of exposure is most unlikely ever to exceed the MEL, to comply with this duty the employer should undertake a programme of monitoring in accordance with regulation 10. This will enable him to show, if it is the case, that the MEL is not normally exceeded, that is, that an occasional result above the MEL is without real significance and is not indicative of a failure to maintain adequate control.

Some substances measured in Schedule 1 of the COSHH Regulations have been assigned short-term MELs i.e. a 10-minute reference period. These substances give rise to acute effects and the purpose of limits of this kind is to render insignificant the risks to health resulting from brief exposure to the substance. For this reason, short-term exposure limits should never be exceeded.

In determining the extent to which it is reasonably practicable to reduce exposure further below the MEL, as required by regulation 7(4), the nature of the risk presented by the substance in question should be weighed against the cost and the effort involved in taking measures to reduce the risk.

Occupational exposure standards (OESs)

An OES is the concentration of an airborne substance, averaged over a reference period, at which, according to current knowledge, there is no evidence that it is likely to be injurious to employees if they are exposed by inhalation day after day to that concentration and which is specified in a list approved by the HSC.

OESs are approved by the HSC following a consideration of the often limited available scientific data by the Working Group on the Assessment of Toxic Chemicals (WATCH).

For a substance which has been assigned an OES, exposure by inhalation should be reduced to that standard. If exposure by inhalation exceeds the OES, then control will still be deemed to be adequate provided that the employer has identified why the OES has been exceeded, and is taking appropriate steps to comply with the OES as soon as is reasonably practicable. In such a case, the employer's objective must be to reduce exposure to the OES, but the final achievement of this objective may take some time. Factors which need to be considered in determining the urgency of the necessary action include the extent and cost of the required measures in relation to the nature and degree of the exposure involved.

LONG-TERM AND SHORT-TERM EXPOSURE LIMITS

Substances hazardous to health may cause adverse effects e.g. irritation of the skin, eyes and lungs, narcosis or even death after short-term exposure, or via long-term exposure through accumulation of substances in the body or through the gradual development of increased risk of disease with each contact.

It is important to control exposure so as to avoid both short-term and long-term effects. Two types of exposure limit are therefore listed in Guidance Note EH40.

The **long-term exposure limit** (**LTEL**) is concerned with the total intake over long periods and is therefore appropriate for protecting against the effects of long-term exposure.

The **short-term exposure limit** (**STEL**) is aimed primarily at avoiding acute effects, or at least reducing the risk of the occurrence. Specific STELs are listed for those substances for which there is evidence of a risk of acute effects occurring as a result of brief exposures.

For those substances for which no STEL is listed, it is recommended that a figure of three times the LTEL averaged over a 10-minute period be used as a guideline for controlling exposure to short-term excursions.

Both LTELs and STELs are expressed as airborne concentrations averaged over a specified reference period. The period for the LTEL is normally eight hours: when a different period is used, this is stated. The averaging period for a STEL is normally 10 minutes, such a limit applying to any 10-minute period throughout the working shift.

'Skin' annotation

Certain substances listed in Guidance Note EH40 carry the skin annotation (Sk). This implies that the substance can be absorbed through the skin. This fact is important when undertaking health risk assessments under the COSHH Regulations.

Examples of MELs and OESs

Table 5.1 gives some examples of substances listed in Guidance Note EH40.

TABLE 5.1 Examples of occupational exposure limits

	Formula	*LTEL*		*STEL*		*Note*
		ppm	mg/m³	ppm	mg/m³	
Maximum exposure limits						
Acrylonitrile	$CH_2=CHCN$	2	4	–	–	Skin
Carbon disulphide	CS_2	10	30	–	–	Skin
Isocyanates		–	0.02	–	0.07	–
Trichlorethylene	$CC1_2=CC1_2$	100	535	150	802	Skin
Occupational exposure standards						
Ammonia	NH_3	25	18	35	27	
Sulphur dioxide	SO_2	2	5	5	13	
Carbon monoxide	CO	50	55	300	330	
Disulphur decafluoride	S_2F_{10}	0.025	0.25	0.075	0.75	
Mercury and compounds (except mercury alkyls)	Hg	–	0.05	0	0.15	

Guidance Note EH40 and the COSHH Regulations

Guidance Note EH40 is an extremely significant document in the interpretation and implementation of the COSHH Regulations. Along with the Approved Codes of Practice (ACOPs) and other documentation issued with the regulations, regular reference should be made to the Guidance Note in activities directed at securing compliance with regulations 6 to 12 of the regulations.

PRINCIPLES OF TOXICOLOGICAL ASSESSMENT AND INVESTIGATION

Toxicological assessment refers to the collection, assembly and evaluation of data on a potentially toxic substances and the conditions of its use, in order to determine the danger to human health, systems for preventing or controlling the danger, the detection and treatment of overexposure and, where such information is insufficient, the need for further investigation. It is closely related to the duties of manufacturers and importers of substances used at work under HSWA, section 6, to ensure the safety of their products, to undertake testing and examination, and to provide adequate information for the user to ensure safe working. These duties are further extended in the Notification of New Substances Regulations 1982 and the Chemicals (Hazard Information and Packaging for Supply) (CHIP) Regulations 1994 and ACOPS accompanying these regulations.

In assessing toxic hazards, and in order to undertake a health risk assessment under the COSHH Regulations, the following basic information is required:

- the name of the substance including any synonyms
- a physical or chemical description of the substance
- information on potential exposure situations
- details of occupational exposure limits
- general toxicological aspects, such as:
  - — routes of entry into the body
  - — the mode of action in or on the body
  - — signs and symptoms
  - — diagnostic tests
  - — treatment
  - — disability potential.

Moreover, under the Notification of New Substances Regulations 1982 specific duties are placed on the manufacturers and importers of substances with regard to testing and notification, announcement procedures, submission of information to the 'competent authority', the form, content and time of notification and procedures for further notification.

Safety data sheets

One of the main problems of toxicological assessment has been the lack of consistency of information given by manufacturers and importers in their safety data sheets. Schedule 5 of the CHIP Regulations lists the following obligatory headings which should be incorporated in a safety data sheet.

1 Identification of the substance/preparation and company
2 Composition/information on ingredients

3 Hazards identification
4 First aid measures
5 Fire-fighting measures
6 Accidental release measures
7 Handling and storage
8 Exposure controls/personal protection
9 Physical and chemical properties
10 Stability and reactivity
11 Toxicological information
12 Ecological information
13 Disposal considerations
14 Transport information
15 Regulatory information
16 Other information.

It is incumbent on the person responsible for the supply or preparation of the substance to supply information specified under the above headings. The safety data sheet should be dated. Detailed guidance on the information to be incorporated in safety data sheets is provided in the ACOP 'Safety Data Sheets for Substances and Preparations Dangerous for Supply: Guidance on regulation 6 of the Chemicals (Hazard Information and Packaging for Supply) Regulations 1994 (HSC)'.

MAIN METHODS OF RECOGNISING HEALTH HAZARDS

Experimental toxicological data

Experimental toxicology is concerned with the study of the effects of exposure to toxic substances. This may entail exposure of laboratory animals, such as rabbits and rats, to controlled doses of toxic substances, measuring the effects on target organs and systems, and comparing same with control animals who have not received these doses.

The terms LD_{50} and LC_{50} are commonly used.

LD_{50} This is the lethal dose for 50 per cent of the test population usually expressed as mg/kg. It is a measure of acute toxicity being the dose of a substance expected to kill 50 per cent of a population of test animals.

LC_{50} This is the lethal concentration for 50 per cent of the test population usually expressed as mg/1 or mg/m^3. It is a measure of acute toxicity being the concentration of a substance in air expected to kill 50 per cent of a population of test animals exposed for a specified period.

As such, both these terms are used as a measure of relative toxicity in relation to human beings.

Epidemiological studies

Epidemiology is the study of the distribution of diseases, such as cholera and, more recently, Legionnaire's disease, in different groups of people. Such studies are important in establishing the causes and effects of disease.

Various criteria must be taken into account in order to establish a specific or direct relationship between the cause of a disease and its effects. These include:

- *Strength* the relative incidence of the disease in exposed and unexposed groups.
- *Consistency* observation of the disease at different locations and times by groups of observers.
- *Specificity* where studies may establish an association between cause and effect which is limited to specific workers and to a certain disease.
- *Biological gradient* is a situation where an increase in the dose is directly related to an increase in the incidence of the disease; this fact may be used to explain the causation.
- *Biological plausibility* the relationship between cause and effect should not conflict with well-established facts as to the aetiology of the disease.
- *Analogy* certain substances or agents may create similar health effects; this fact should be known by researchers.
- *Preventive action* where preventive action has been successful in reducing the incidence of the disease, then it is fairly certain that the cause was correctly established.

Epidemiological studies tend to take two forms – cohort studies and case-controlled studies.

Retrospective studies are commonly used to determine whether there is an association, for example, between exposure to a certain chemical substance and a certain level of diagnosed cases of dermatitis. They use two groups or cohorts of people, those who have been exposed to the substance and an unexposed (control) group.

Cohort studies endeavour to establish a direct relationship between the causes and effect i.e. X number of diagnosed cases.

Case-controlled studies, on the other hand, might be used to investigate the frequency of chemical workers who suffer various levels of dermatitis against a control group comprising members of the public. These studies tend to be less effective in establishing cause and effect, however.

Information from data, etc.

Various forms of data are produced by organisations such as the Department of Health, HSE and local authorities with regard to, for instance, the number of diagnosed cases per year of particular diseases. These may be expressed in terms of incidence rate i.e. the total number of cases per 1,000 employees or members of the population.

Such information may also be presented as bar charts showing the number of cases on a year to year basis, or as pie charts showing the percentage of occupational diseases of various types diagnosed.

THE ROLE OF ACTS AND WATCH

The Advisory Committee on Toxic Substances (ACTS)

This committee operates under the auspices of the HSC and includes representatives of the Confederation of British Industry, the Trades Union Congress, local authorities, government departments and independent experts. The committee provides advice to the HSC on the toxicological aspects of existing and new substances.

The Working Group on the Assessment of Toxic Chemicals (WATCH)

In the UK, values for occupational exposure standards are approved by the HSC following consideration of the often limited scientific evidence by WATCH, which also subjects all OEL values to a regular review.

SAFE HANDLING OF TOXIC SUBSTANCES

To summarise, a number of general principles can be considered in the safe handling of toxic substances. Assuming that substitution as a control strategy has been considered and found impracticable then, in order of merit, the following practices and procedures should be followed:

- use in diluted form wherever possible
- only limited quantities should be used or stored at any one time; large quantities should be stored in a purpose-built bulk chemical store
- in certain cases, containment of a specific area may be necessary; consider safe venting and drainage requirements
- eliminate handling and dispensing from bulk; in certain processes, the use of automatic systems e.g. cleaning-in-place (CIP) systems may be possible
- provide adequate LEV systems, which are subject to regular examination and test
- in certain cases, separation of substances may be feasible e.g. acids from alkalis
- personal protective equipment as an extra means, not sole means, of protection.

Personal protective equipment and associated facilities

The following matters should be considered in the selection and use of various items of PPE to ensure appropriate protection from toxic substances:

- *Protective clothing* one-piece overalls, aprons, gloves for specific duties; frequency of laundering important.
- *Footwear* wellington boots with steel toe-caps and insoles; safety boots and gaiters.
- *Eye protection* spectacles, goggles or visors according to the extent of handling and degree of risk; specific recommendations for various jobs where necessary.
- *Respiratory protection* full face mask, ori-nasal, air-fed types; these should feature in a COSHH health risk assessment.
- *Emergency provisions* first aid, emergency showers, eye wash bottles; decontamination chemicals and procedures.
- *Welfare amenity provisions* adequate provision for sanitation, hand washing, showers, clothing storage; use of barrier creams and skin cleansers.

Environmental and biological monitoring

One definition of the term stress is 'the common response to environmental change'. Stress in the working environment may be created in a number of ways as a result of, for instance, the installation of noisy plant and equipment. Stress factors may also take many forms, for example, extremes of temperature, poor levels of lighting and ventilation or the presence of hazardous dusts, gases, vapours and bacteria, all of which have a detrimental effect on the health of persons exposed to such stressors.

6

In order to prevent ill health arising as a result of poor working environments, employers have a duty under section 2(2)(e) of HSWA to provide and maintain for their employees a working environment that is, so far as is reasonably practicable, safe, without risks to health, and adequate as regards facilities and arrangements for their welfare at work.

Environmental monitoring is concerned with the identification of these physical, chemical and biological stressors, such as noise, dust, toxic gases and harmful bacteria, in the workplace environment. As such, it is an important measure in identifying individual and group exposure to these stressors and in ensuring compliance with the COSHH Regulations and other regulations, such as the Noise at Work Regulations 1989 and the Ionising Radiations Regulations 1985.

Biological monitoring, on the other hand, is a regular measuring activity where selected validated indicators of the uptake of toxic substances in the human body are determined in order to prevent health impairment. This form of monitoring could entail examination of, for example, blood, urine, saliva and expired air. Biological monitoring commonly features in the health or medical surveillance of persons exposed to hazardous environments under the COSHH Regulations and the Control of Lead at Work Regulations 1980.

Environmental monitoring, including air monitoring, is an important aspect of occupational hygiene practice.

Occupational hygiene

Occupational hygiene is defined as being concerned with the identification, measurement and control of contaminants and other phenomena, such as noise and radiation, which would have otherwise unacceptable adverse effects on the health of people exposed to them. (Annual Report of HM Chief Inspector of Factories, 1973.)

It is concerned with the monitoring and control of the working environment to ensure that contaminants are kept to as low a level as is reasonably practicable and, in all cases, to a level that is above the appropriate hygiene standard, that is, below an identified exposure limit.

OCCUPATIONAL HYGIENE PRACTICE

Wherever environmental contamination is present, or suspected of being present, the following sequence of operations is necessary:

1 identification of the stressor
2 measurement and evaluation of same against formal hygiene standards
3 selection of an appropriate prevention or control strategy
4 implementation of the selected strategy
5 monitoring on a frequent basis to ensure the hygiene standard is maintained.

These features of occupational hygiene practice are discussed below.

Recognition and identification

The recognition and subsequent identification of the specific contaminant e.g. dust, fumes, gas, vapour, mist, virus, sound pressure level, is the first stage in the sequence. A number of spot check devices are used, such as detector stain tubes for gases, coated slides for dust or, in the case of noise, a sound pressure level meter.

Measurement

Once the contaminant has been identified, it is necessary to measure the extent of the contamination. A wide range of equipment is available for the accurate measurement of environmental contamination. (See 'Measurement techniques' later in this chapter.)

Evaluation

Evaluation is an important part of the procedure. Measured levels of contamination must be compared with current hygiene standards, always assuming there is such a standard applicable to the substance in question, such as occupational exposure limits (maximum exposure limits and occupational exposure standards). In addition, the duration and frequency of exposure to the contaminant must be taken into account.

Following a comprehensive evaluation, a decision must be made as to the actual degree of risk to those exposed. This degree of risk will determine the prevention or control strategy applied, such as the installation of local exhaust ventilation equipment, soundproof enclosure, etc.

Prevention or control of exposure

The COSHH Regulations require that wherever a risk to health is identified, exposure must either be prevented or controlled. Prevention or control of exposure is the fourth stage of occupational hygiene practice. In extreme cases it may be necessary to prohibit the use of a toxic substance; in less severe situations the risk to the health of workers exposed may be eliminated by, perhaps, isolating the source of contamination or changing a process to afford better operator protection, coupled with the provision of information, instruction, training and effective supervision.

Monitoring

Once the prevention or control strategy has been installed, it is management's responsibility to ensure that these measures are maintained at all times. This may entail regular maintenance, examination and testing of exhaust ventilation equipment and/or some form of health surveillance.

AIR MONITORING AND MEASUREMENT TECHNIQUES

The identification and measurement of airborne particulates – dusts, gases, vapours, etc. involves the taking of air samples for measurement purposes (see Fig 6.1).

HSE Guidance Note EH42 'Monitoring Strategies for Toxic Substances' gives practical advice to employers on how they should conduct investigations into the nature, extent and control of exposure to substances hazardous to health which are present in the air of the workplace.

Air sampling

Air sampling can be undertaken on either a short-term or long-term basis. Long-term sampling may entail the use of personal dosimeters (personal sampling) and/or static sampling systems.

Short-term sampling techniques (grab sampling, snap sampling)

Short-term sampling implies taking an immediate sample of air and, in most cases, passing it through a particular chemical agent which responds to the contaminant being monitored.

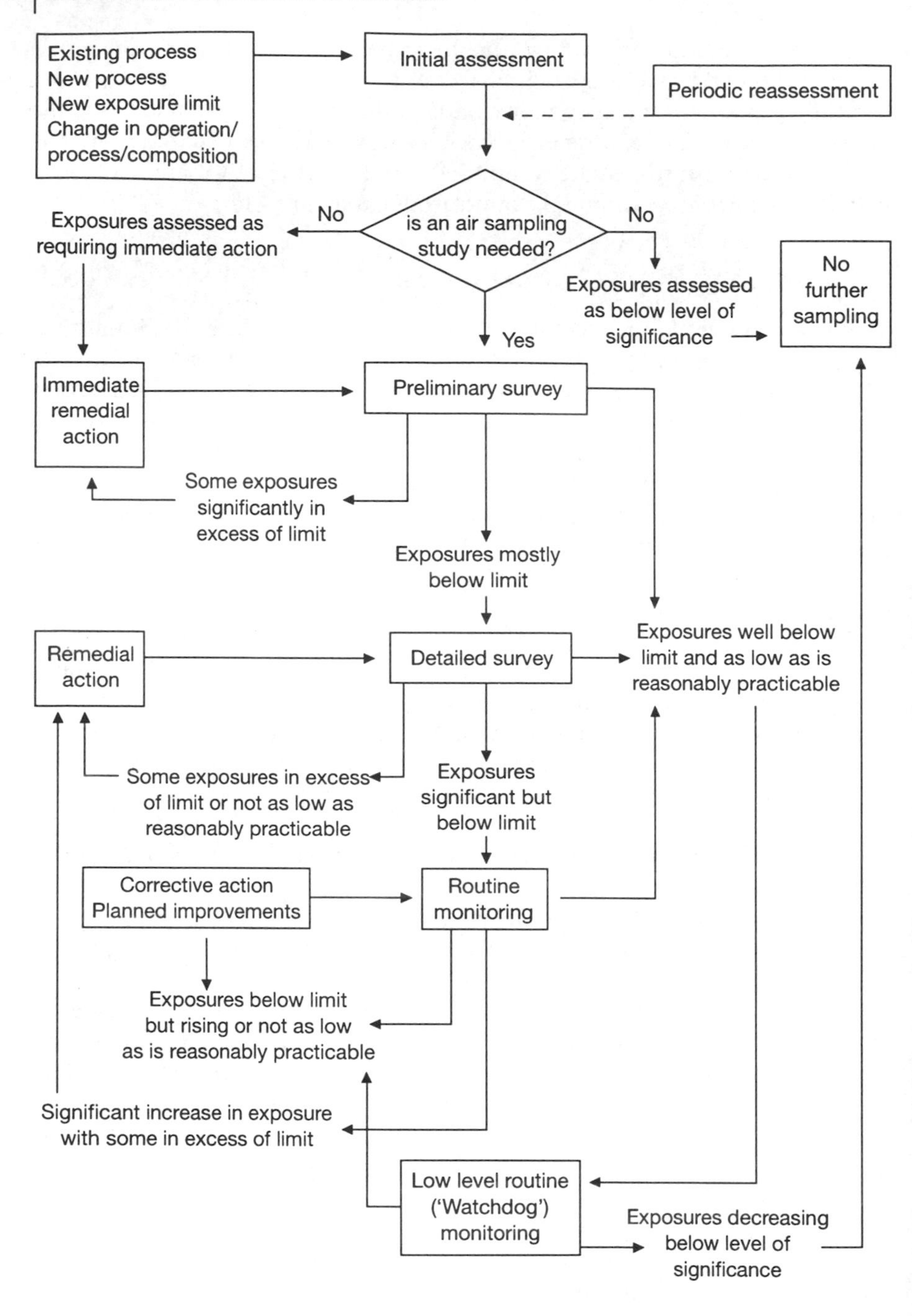

• **FIG 6.1 Monitoring strategies flow diagram**

One form of simple device for the measurement of concentrations of gases and vapours is the hand pump and bellows device (multi-gas detector – see Fig 6.2) which incorporates a specific detector tube. The detector tube is a glass tube, sealed at both ends, and filled with porous granules of an inert material, such as silica gel. The granules are impregnated with a chemical agent which changes colour in the presence of the contaminant gas. In order to detect and measure this gas, the ends are broken off the tube and the tube inserted into the tube holder of the sampling pump. The correct volume of air is drawn through the detector tube (see Fig 6.3) according to the number of times the bellows are depressed and the resultant colour stain indicates the presence of the gas. The actual concentration can be determined either by the length of the stain which increases proportionally with the concentration, or by comparing the intensity of the colour with a prepared standard. The sampling pump is small, light and hand-operated. This combination of pump and detector tube provides a convenient method of on-the-spot evaluation of atmospheric contamination.

The hand pump and bellows device is a useful instrument for the early detection of atmospheric contamination. Users do not require extensive training and it can be used relatively cheaply in random sampling exercises or routine day-to-day monitoring. However, its accuracy must always be

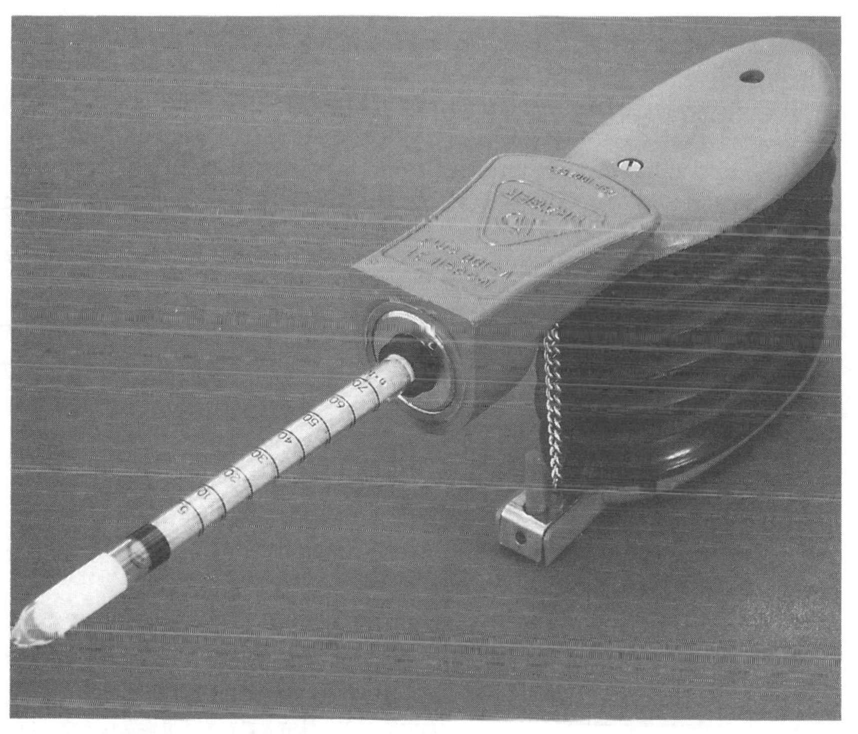

• **FIG 6.2 Multi-gas detector**
Reproduced by courtesy of Draeger Safety

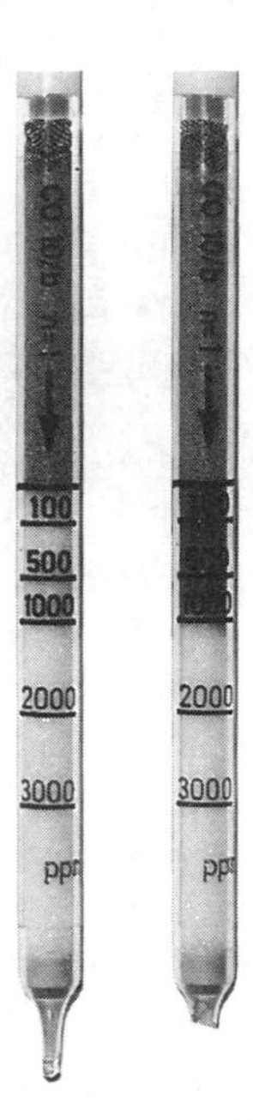

• **FIG 6.3 Gas and vapour detector tubes**
Reproduced by courtesy of Draeger Safety

suspect owing to the problem of cross-sensitivity of detector tubes, tubes being used that have gone past the expiry date or even the operator failing to give the appropriate number of pump strokes when operating the bellows. Therefore, when frequent monitoring must be undertaken, an automatic system should be used. Furthermore, normal stain tubes will only give 'point in time' results and, dependent on circumstances, may not give an adequate picture of major fluctuations across a working shift.

Long-term sampling techniques

Instruments which carry out long-term sampling are, broadly, of two types, personal samplers and static samplers.

Personal sampling instruments

These may take a number of forms e.g. gas monitoring badges, filtration devices and impingers. The principal purpose of the use of personal sampling instruments is that of measuring the specific exposure of individuals, particularly where they may not be exposed to the environmental stressor on a continuous basis.

With gas monitoring badges, the worker wears a badge containing a solid sorbent or a chemically impregnated carrier, and the air sample comes into contact with the badge by diffusion. The results can be read directly by a colour change or determined by analytical instruments.

Filtration devices comprise a low-flow or constant-flow sample pump, which is motor-operated from a rechargeable battery, and a sampling head, which incorporates a specific filter, attached close to the operator's breathing zone (see Fig 6.4). Many dusts, mists, etc. can be collected by passing a known volume of air through a filter, the pore diameter of which is selected to remove the chemical hazard completely. The quantity of dangerous material collected may be determined gravimetrically (weighing the filter before and after collection) or by solvent extraction and analysis by gas chromatography, atomic absorption, etc. Filtration methods are particularly useful in the sampling of large particles or aggregates of particles.

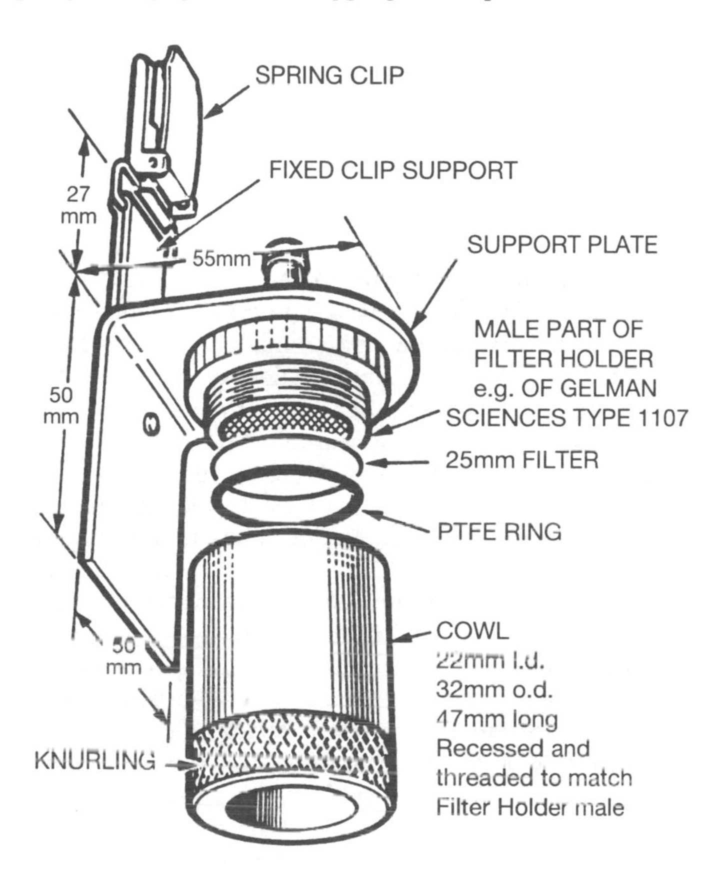

• **FIG 6.4 Sketch of a suitable filter holder**

The impinger method is used for collecting such chemical compounds as acetic anhydride, hydrogen, chloride, etc. The impinger is a specially designed glass bubble tube. A known volume of air is bubbled through the impinger containing a liquid medium chosen to react chemically to, or physically to dissolve, the contaminant. The liquid is then analysed by gas chromatography, spectrophotometry, etc. An impinger operates in conjunction with a constant flow sample pump. It may be mounted on the side of a sample pump, which is worn on the operator's belt, or in a holster near the breathing zone.

Another method of personal sampling involves the aid of a sorbent tube, which is a method for collecting a large percentage of the hazardous chemical vapours in a work environment. A glass sample tube is used, normally filled with two layers of a solid absorbent capable of completely removing chemicals from the air. The tube has breakable end tips. To collect a sample, the end tips are broken and a known volume of air is drawn through the tube. Airborne chemicals are trapped by the first absorbent layer, with the back-up layer assuring removal of all chemicals from the air. The tube is then sealed with a push-on cap prior to analysis. The tube may be inserted into a tube holder, located in the breathing zone of the operator, which is connected to a constant flow pump, or directed into the pump by means of a short extension piece.

Static samplers

These devices are stationed in the working area. They sample continuously over the length of a shift, or longer period if necessary. Mains or battery operated pumps are used. They sample contaminants which may be present in the general atmosphere in very small quantities but which nevertheless may be dangerous. Such pumps can handle large or small quantities of air per minute and pass it through a variety of sampling devices. For obtaining samples of harmful dusts, such as asbestos or silica, filters of various porosity ranges are used. When the particle size of the dust is significant, size-selective filters are used. Moreover, when the substance to be sampled is volatile e.g. a solvent vapour, or must be analysed in solution, the contaminant must be trapped on to a suitable medium, such as activated charcoal, or in an absorbing liquid, for subsequent laboratory analysis.

Long-term stain detector tubes are available for this purpose. The device is connected to a constant-flow sample pump and air is drawn through the tube at a steady rate. Examination of the detector tube at the termination of the sampling period will indicate the amount of contamination absorbed, which is directly related to the average level of contamination present over the period.

A number of direct monitoring devices are also available for the detection and measurement of gases and vapours. They operate on several different principles e.g. infra-red absorption, and give an instant read-out on a chart, meter or display. In many cases they can be linked to an alarm device which sounds once a particular concentration of gas or vapour reaches a predetermined level.

Limitations in the information collected from air sampling exercises

Any air sampling exercise is affected by a number of factors which can influence the result of such an exercise. Many variables affect airborne concentrations at a particular point in time and, on this basis, it may be necessary to undertake several sampling exercises at different times in order to obtain an accurate assessment of environmental contamination and its potential for harm.

The occupational hygiene process is summarised in Fig 6.5.

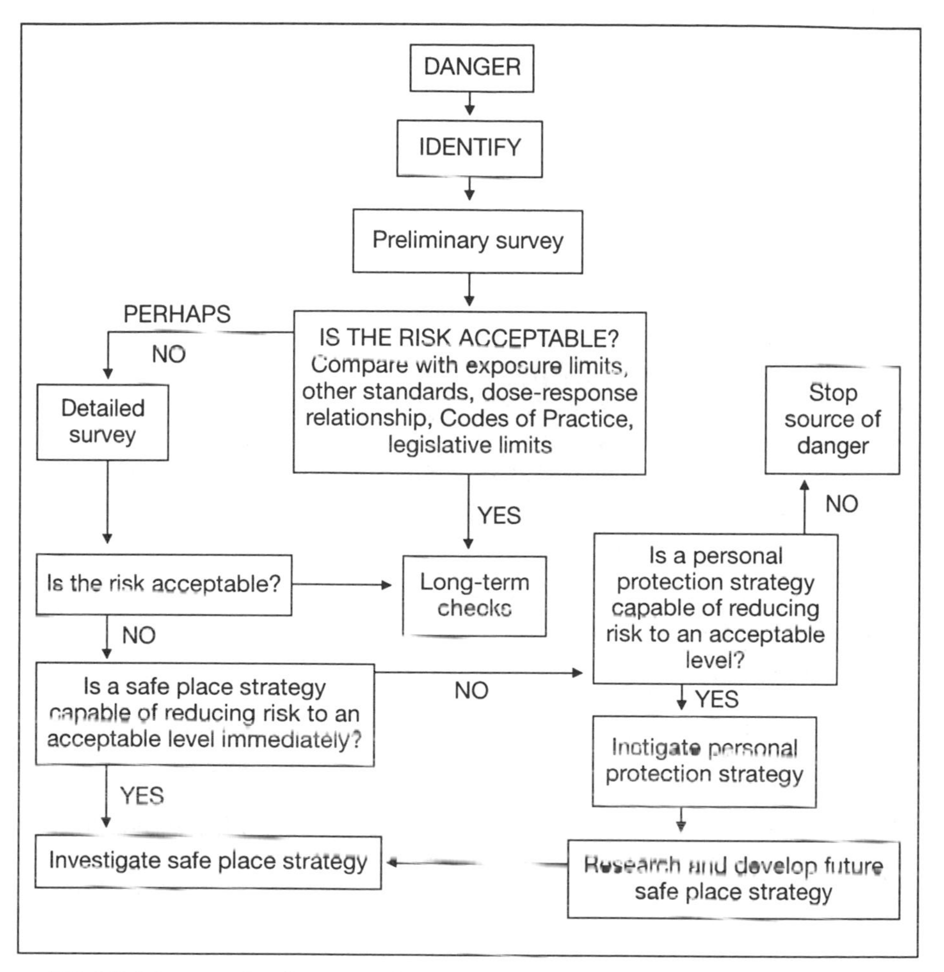

• **FIG 6.5 The standard approach to occupational health hazards**

These variables include:

- the type, position and rate of release of the contaminant from each source
- the effects of air currents and turbulence in the workplace which can affect the dispersion or mixing of contaminants
- the effects of external conditions, in particular wind speed and direction, air temperature and humidity

- the fact that workers may not, by virtue of variations in their job movements, operations and activities, be permanently exposed to the contaminant for the total duration of the sampling period.

This last mentioned point is important and the following variations need consideration:

- variations in shift patterns and the average exposure of individual workers
- variations associated with the type and nature of different processes in terms of ambient environmental conditions e.g. temperature
- variations in the contaminant concentration in the breathing zones of operators over the duration of a shift
- variations in individual exposure levels, even when working in the same place, doing the same work and on the same shift.

BIOLOGICAL MONITORING

Biological monitoring was defined at the commencement of this chapter as 'a regular measuring activity where selected validated indicators of the uptake of toxic substances in the human body are determined in order to prevent health impairment'. It is undertaken through the determination of the effects certain substances produce on biological samples of exposed individuals and these determinations are used as biological indicators.

Dose-effect relationship

The evaluation of the relationship between the dose of an offending agent and the effect on the human body of that specific dose is based on analysis of the degree of association existing between an indicator of dose e.g. blood, urine, saliva, faeces, and an indicator of effect on the body e.g. loss of consciousness, lacrimation, respiratory response. The study of the dose-effect relationship identifies at which concentrations of toxic substance the indicator of effect exceeds the values currently accepted as 'normal'.

Dose-response relationship

This term is discussed in Chapter 5. Since not all individuals in a group react in the same manner to exposure to a contaminant, it is necessary to study how the group responds by evaluating the appearance of the effect compared to internal dose. This is what is meant by 'response', which is the percentage of subjects in the group who show a specific quantitative variation of an indicator of each dose level.

BIOLOGICAL INDICATORS

The biological samples where the indicators may be determined consist of:

- blood, urine, saliva, sweat, faeces
- hair, nails
- expired air.

With biological monitoring, information can be obtained that would not otherwise be available i.e:

- on the evaluation of absorption and/or exposure over a prolonged period of time
- on the amount of substance absorbed as a result of movements within the working environment or of accidental causes, which often cannot be checked
- on the amount absorbed by the organism via various routes of entry
- on the evaluation of overall exposure, as the sum of different sources of contamination, which may also exist outside the working environment
- on the amount absorbed by the subject, taken as an individual, as related not only to his workplace, but taking into account climatic factors, specific physical resistance, age, sex, individual genetic characteristics, etc.
- on whether the subject has been exposed to a risk which would not be proven in any other way and, in some cases, when.

The indicators of internal dose can be further divided into:

- *True indicators of dose* i.e. capable of indicating the quantity of the substance at the sites where it exerts its effect
- *Indicators of exposure* that can provide an indirect estimate of the degree of exposure, since the levels of the substances in the biological samples closely correlate with levels of environmental pollution.
- *Indicators of accumulation* that can provide an evaluation of the concentration of the substance in organs and/or tissues from which the substance, once deposited, is slowly released.

Biological exposure indices (BEIs)

BEIs are set by the American Conference of Government Industrial Hygienists (ACGIH) to reflect the average body fluid concentration of a toxic substance or its metabolite found in workers exposed at the equivalent threshold limit value (TLV).

Biological/tolerance values

These values are established in Germany and are defined as 'the maximum permissible quantity of a chemical compound, or its metabolites or any deviation from norm of biological parameters induced by these substances in exposed humans'. According to current knowledge, these conditions do not impair the health of employees, even if the exposure is repeated and of long duration.

The German values are thus health-based values and are set as acceptable upper limits, whereas the ACGIH levels are only for the guidance of occupational health professionals. In a group of workers exposed at a given TLV there will be a typical biological spread of the concentration of toxic substances or their metabolities in biological fluids.

The BEI is, therefore, not an acceptable upper limit and should only be considered as a guide to the occupational physician or hygienist. In fact, all it may indicate is the average concentration to be expected in a worker exposed at the TLV.

Detection by biological monitoring

Environmental monitoring and control of the absorption of some toxic materials into the body is not possible without monitoring the total uptake of the material in the body. Biological monitoring is a valuable additional tool available to occupational health professionals working in this area.

Furthermore, environmental monitoring of breathing zone air may not be a good guide to the levels of contaminant absorbed by the body in the long term because of the variations in:

- breathing rate (work rate)
- metabolic exit rate
- absorption rate in the lungs.

Analysis of the following can be used to gauge atmospheric exposure levels (see Table 6.1).

TABLE 6.1 Hazardous substances that can be revealed by medical analysis

Substance	*Technique*
Benzene	Phenol in urine; benzene in breath
Inorganic lead	Lead in blood/urine; coproporphyrin III in urine
Elemental mercury/ mercury	Mercury in urine; protein in urine
Methyl mercury	In faeces
Arsenic	In urine, hair, nails
Cadmium	In blood and urine
Trichlorethylene	In urine as trichloracetic acid
Organo-phosphorus compounds	Cholinesterase in blood/urine; nerve conduction velocity; electromyography

NOISE MONITORING

Basic theory of noise measurement

A sound pressure level meter measures sound intensity on a comparative basis. The range of intensities to which the ear responds, however, is enormous, from the **threshold of hearing** to the **threshold of pain**. For example, at a frequency of 1,000 Hz the threshold of pain is 100,000,000,000,000 (10^{14}) times more intense than the threshold of hearing, where sound is just discernible. It is clearly difficult to express such ratios on a simple arithmetic scale, so a logarithmic scale is used. The ratio would therefore be expressed as:

$$\log_{10} \frac{10^{14}}{1} \text{ or } 14, \text{ rather than } \frac{10^{14}}{1}$$

The unit used is the **bel**. Thus:

1 bel is $\log_{10}10^1$ (a tenfold change in intensity)

2 bels is $\log_{10}10^2$ (a hundredfold change in intensity)

and so on.

6

The bel, however, is a very large unit, so it is further split into tenths, called **decibels (dB)**: 1 bel equals 10 decibels.
For example:

$$10 \log_{10}10^{14} \text{ equals } 140 \text{ dB}$$

Thus 1 decibel equals a change in intensity of 1.26 times, since $10^{1/10}$ is 1.26 (or 1.26^{10} is 10). Also a change in intensity of 3dB = 1.26^3 = 2, so that doubling the intensity of a sound gives an increase of 3 dB.

If there are two sounds of intensities I_1 and I_2 and they differ by n dB, then:

$$n = 10 \log_{10} \frac{I_1}{I_2}$$

It is normal practice to relate intensity to a standard reference level, so that:

$$n = 10 \log_{10} \frac{I_1}{I_0}$$

and I_0 is taken as 10^{-12} watts per square metre.

However, as intensity is proportional to pressure squared:

$$n = 10 \log_{10} \frac{P^2}{P_0^2}$$

$$= 10 \log_{10} \left[\frac{P}{P_0}\right]^2$$

$$\text{or } n = 20 \log \frac{P}{P_0} \text{ dB}$$

where P is the standard reference level of 2 x 10^{-5} newtons per square metre (**pascals**) and n is the sound pressure level in dB. Pressure is the easiest quantity to measure, hence the use of dB sound pressure level. The standard reference level of 2 x 10^{-5} N/m^2 is chosen since it is the average threshold of audibility at 1,000 Hz (i.e. it is 0 dB).

Note: Under the SI system, sound pressure is expressed in pascals. A **pascal** is a unit of pressure corresponding to a force of one newton acting uniformly upon an area of 1 square metre. Hence 1 Pa = 1 N/m^2.

The use of a logarithmic scale in sound measured has a further advantage, because the evaluation of intensities is simplified by the replacement of multiplication with addition and of division with subtraction. Furthermore, the response of the ear tends to follow a logarithmic scale.

Addition of decibels

The addition of decibels is carried out on a ratio basis, rather than an arithmetic one, and Table 6.2 may be used to simplify the procedure. To add two sound pressure levels, take the difference between the two levels and add the corresponding figure in the right-hand column to the higher sound pressure level.

TABLE 6.2 Addition of decibels

Difference (dB)	*Add to higher (dB)*
0.0–0.5	3.0
1.0–1.5	2.5
2.0–3.0	2.0
3.5–4.5	1.5
5.0–7.0	0.5
Over 12.0	0.0

Other units used in noise measurement

Phons

The human ear does not respond equally to all frequencies. Sounds of different frequency at a constant sound pressure level do not evoke equal loudness sensations. This phenomenon is linear with neither amplitude nor frequency, and 'loudness level' is measured in phons, the sound being compared again to a standard reference signal of 1,000 Hz. The loudness level in phons of any sound is taken as that which is subjectively as loud as a 1,000 Hz tone of known level. 0 phon is 0 dB at 1,000 Hz. 50 phons is the loudness of any tone which is as loud as a 1,000 Hz tone of 50 dB. This can be demonstrated by **equal loudness curves** shown in Fig 6.6. Maximum sensitivity occurs between 1 and 5 Hz. The curves are obtained by finding the sound levels at different frequencies which seem equally loud to the listener in comparison with a reference sound at 1 kHz.

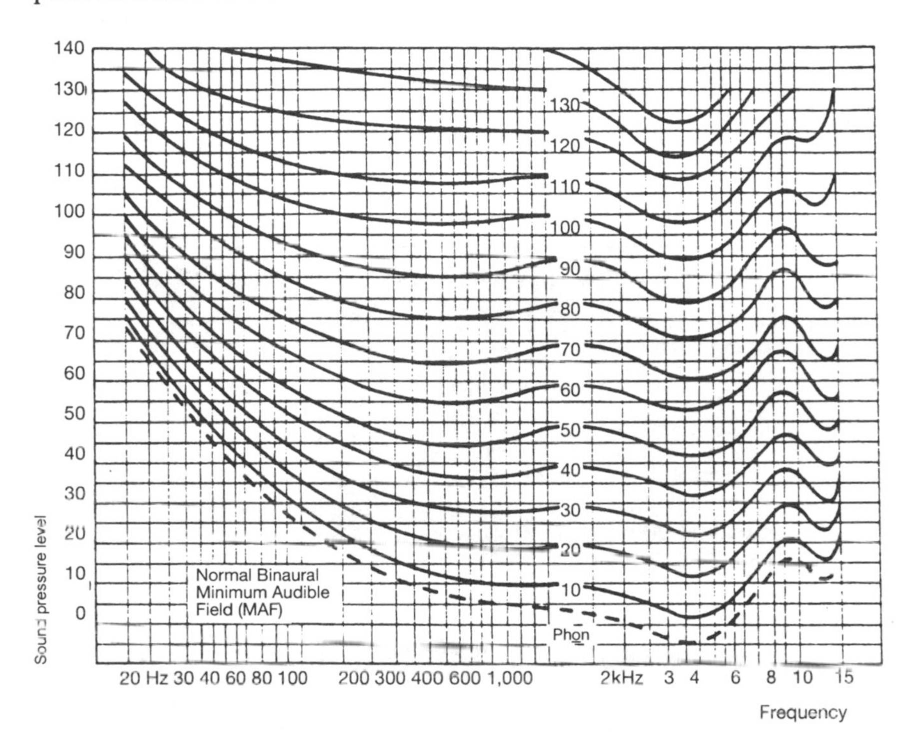

• **FIG 6.6 Equal loudness curves**

Sones

This is a linear unit of loudness on a scale designed to give scale numbers approximately proportional to loudness. The scale is precisely defined by its relation to the phon scale.

OCTAVE BANDS AND OCTAVE BAND ANALYSIS

It is possible to make a single measurement of the overall sound pressure of the entire range of audible frequencies but this measurement, if taken in linear decibels, is of limited use since the ear is more sensitive to some frequencies than others. Use of the 'A' weighted decibel scale (see below) provides a reasonable means of assessing likely risk to hearing but a knowledge of the way in which the sound is distributed throughout the frequency spectrum provides a much more accurate picture. This can be obtained by dividing the noise into octave bands and measuring the sound pressure level at the centre frequency of each band. (An octave represents a doubling of frequency, so that the range 90–180 Hz is one octave, as is the range, 1,400–2,800 Hz.)

The octave bands are usually identified by their geometric centre frequencies. For example, the geometric centre frequency of the octave 90–180 Hz is approximately 125 Hz. The standard range of octave bands has the geometric centre frequencies as shown in Table 6.3. Octave band analysis is used for assessing risk of noise-induced hearing loss and in the specification of certain forms and types of hearing protection. It is also used in the diagnosis of machinery noise and in the selection of noise attenuation methods.

TABLE 6.3 Standard range of octave bands

Limits of band (Hz)	*Geometric centre frequency (Hz)*
45–90	63
90–180	125
180–355	250
355–710	500
710–1,400	1,000
1,400–2,800	2,000
2,800–5,600	3,000
5,600–11,200	8,000

THE SOUND PRESSURE LEVEL METER

A sound pressure level meter is an instrument which measures linear sound pressure level in the human audiofrequency range unless provided with and set to various weighting networks. The 'A' weighted network gives objective measurements of sound pressure level in accordance with the manner of response of the human ear. A typical mode of operation is shown in Fig 6.7.

The microphone senses the air pressure fluctuations and converts mechanical vibration to an electrical signal containing amplitude and frequency components. The amplifier increases the weak signal from the microphone and incorporates gain adjustments, which enable the instrument to cope with the very wide range of pressure amplitudes which the ear can sense. The sound signal is also available as an output socket so that it may be fed to external instruments such as recorders or noise dosimeters.

Since an accurate response from the sound level meter is necessary, provision is made to calibrate it for accurate results. This is done by the use of a portable acoustic calibrator placed directly over the microphone. The calibrator is basically a miniature audible signal generator giving a pre-

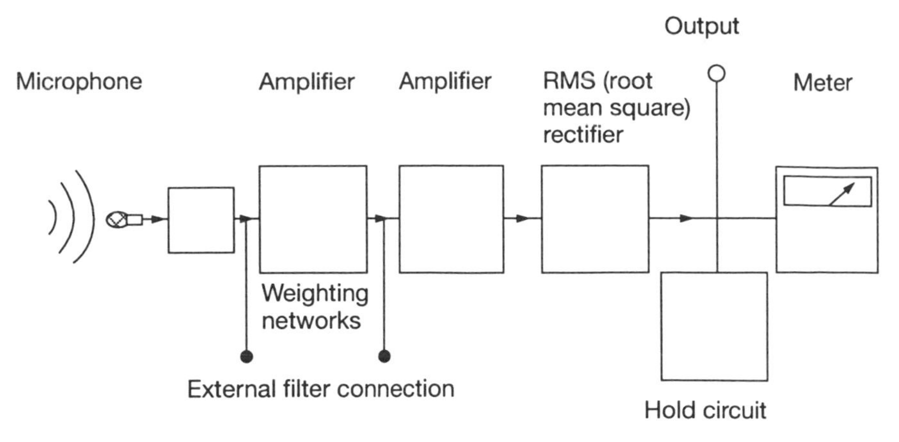

• **FIG 6.7 Sound pressure level meter**

cisely defined sound pressure level to which the sound level meter can be calibrated. Electronic oscillators are more commonly used.

When the sound level fluctuates, the meter needle should follow these variations. However, if the level fluctuates too rapidly, the meter needle may move so erratically that it is impossible to obtain a meaningful reading.

For this reason, two meter responses are used:

- *Fast* this gives a fast reacting indicator response which enables the user to follow and measure noise levels which are not fluctuating so rapidly.
- *Slow* this gives a damped response and helps average out meter fluctuations which would otherwise be impossible to read.

Weighting networks

The sound level meter incorporates electrical circuits known as weighting networks. These provide for various sensitivities to sounds of different frequencies, the original object being to simulate the response characteristics of the human ear at different frequencies. These weighting networks are known as 'A', 'B', 'C' or 'D' weighted decibel scale operating conditions and can be selected on a sound level meter.

'A' scale

The 'A' scale is normally used for industrial noise measurement. This scale makes the instrument more sensitive to the middle range of frequencies, and less sensitive to high and low frequencies, and is the one which most closely approximates to the response of the human ear.

'B' scale

This scale was intended for the measurement of middle range sound pressure levels, between 55 and 85 dB. It is not commonly used as it does not give good correlation to subjective tests of hearing perception.

'C' scale

This gives most sensitivity in low frequencies and is, therefore, of limited use. As with the 'B' scale, it does not give good correlation to subjective tests.

'D' scale

This scale is generally limited to the measurement of aircraft noise and has little or no application in the measurement of industrial noise.

RADIATION MONITORING

A number of instruments are available for the measurement and detection of radioactivity.

Geiger-Muller counter

The operation of the instrument is based on the fact that ionising radiations produce electrical charges which can be detected. One 'click' of a Geiger counter corresponds to one atom of radioactivity being detected. The Geiger counter (see Fig 6.8) is an extremely sensitive instrument and can be used for monitoring background radiation to the extent that a dose of one-fiftieth (background) of the maximum permissible dose can be detected.

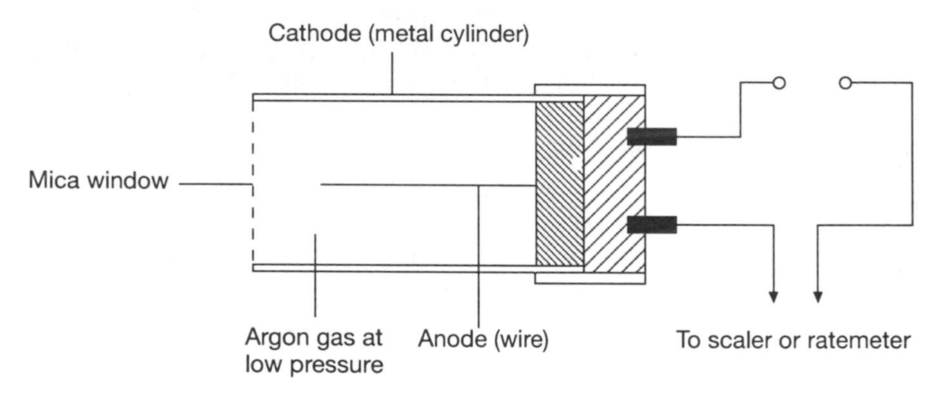

• **FIG 6.8 Geiger counter**

When radiation enters the tube, either through a thin window made of mica, or, if it is very penetrating, through the wall, it creates argon ions and electrons. These are accelerated towards the electrodes and cause more ionisation by colliding with other argon atoms. On reaching the electrodes, the ions produce a current pulse which is amplified and fed either to a scaler or a ratemeter. A scaler counts the pulses and shows the total received in a certain time. A ratemeter has a meter marked in counts per second (or minute) from which the average pulse rate can be read. It usually incorporates a loudspeaker which gives the characteristic click for each pulse. This instrument cannot deal with count rates in excess of 1,000 to 2,000 counts per second. It is very useful, however, for measuring low levels of radiation. It will not generally detect the presence of beta particles due to the relative thickness of the window.

Scintillation counter

This instrument measures radiation intensity and uses a screen of material which emits flashes of light when bombarded with alpha, beta, gamma and/or slow neutron radiation. These light flashes are converted to an electric current which increases as the bombardment increases. The current is amplified and indicated in the same way as the bombardment increases. The current is amplified and indicated in the same way as with the Geiger-Muller counter.

Airborne sampler

As with airborne dust, certain radiations can be sampled on a filter paper using a high-volume air sampler. A known volume of air is drawn through a filter which is removed at the end of the sampling period and scanned for its radioactivity, using a counter.

Film badges

A film sensitive to radiation is housed in a specially designed plastic casing containing windows of various materials which shield certain kinds of radiation, but which allow others to pass through. This allows assessment of the various types of radiation. The device is worn during periods of exposure, one badge generally lasting a week. The film is developed and analysed for the accumulated dose of the various types of radiation, and a permanent record of the worker's personal exposure is produced.

Thermoluminescent personal dosimeter (TLD)

Some materials, such as lithium fluoride, can convert to an 'excited' state when bombarded with ionising radiation. This state is reversed only on the application of heat when the crystals return to normal, but with a measurable emission of light. Thus, a small badge containing these crystals can be used as a dosimeter, since the degree of irradiation can be related to the amount of light produced on heating. An advantage with this type of dosimeter is that it is small and its analysis can be quickly and automatically performed.

Quartz fibre detector (packet electrometer)

This detector consists of a metal cylinder with a loop of quartz fibre in the middle which is charged up to approximately 200 volts, electrical attraction or repulsion taking place between the fibre and the outside case. Ionising radiation will allow the charge to leak away slowly, permitting the loop to move over a scale back to the centre of the cylinder. This instrument will measure down to 5 millirems of dose and is ideal for use in short-term exposure situations e.g. one–two weeks.

Prevention and control strategies

THE LEGAL DUTY TO PREVENT OR CONTROL EXPOSURE

The HSWA clearly defines the duty of the employer to 'provide and maintain a working environment for his employees that is, so far as is reasonably practicable, safe, without risks to health, and adequate as regards facilities and arrangements for their welfare at work'. The COSHH Regulations 1994, the Noise at Work Regulations 1989 and the Workplace (Health, Safety and Welfare) Regulations 1992 greatly reinforce this duty.

Regulation 7(1) of the COSHH Regulations in particular requires every employer to 'ensure that the exposure of his employees to substances hazardous to health is either prevented or, where this is not reasonably practicable, adequately controlled'. Moreover, under regulation 7(2), 'so far as is reasonably practicable, the prevention or adequate control of exposure of employees to a substance hazardous to health shall be secured by measures other than the provision of personal protective equipment'.

In the prevention and control of health risks, many approaches are available, depending upon the severity and nature of the health stressor concerned. The principal prevention and control strategies, together with the various support strategies are outlined below.

PREVENTION STRATEGIES

Prohibition

This is the most extreme form of prevention strategy applicable and commonly forms the basis for a prohibition notice served by an enforcement officer. As a strategy, prohibition characterises much of the legislation relating to, for instance, known carcinogens, as well as other hazards where there is no known form of protection available to those exposed.

Prohibition, therefore, implies a total ban on a particular system of work, substance used at work or the operation of a working practice where the danger level is unacceptable.

Elimination

Reviews of the needs of specific processes often reveal potentially hazardous substances whose use is no longer necessary. Such substances should be eliminated thereby negating the need for a form of control.

Substitution

The substitution of a less toxic material in place of a more highly toxic one is a frequently used strategy. Typical examples are the substitution of toluene for benzene and trichloroethane 1,1,1 in place of carbon tetrachloride.

Wherever possible, substances with an assigned maximum exposure limit (MEL) detailed in Guidance Note EH40 'Occupational Exposure Limits' should be substituted by substances not quoted in the list of MELs. Health risk assessments undertaken to comply with the COSHH Regulations will, in many cases, identify safer substances which can be substituted for the more dangerous ones.

CONTROL STRATEGIES

Containment (enclosure)

This strategy is based on the containment of an offending agent or environmental stressor to prevent its liberation into the working environment. Total enclosure or containment of a process may be possible by the use of bulk tanks and pipework to deliver a liquid directly into a closed production vessel. Complete enclosure is practicable if the materials are in liquid form, used in large quantities, and where the range of materials in use is very limited.

Enclosure may take a number of forms e.g. acoustic enclosures for noisy machinery, dust enclosures, paint spray booths, laboratory fume cupboards.

Isolation (separation)

The physical isolation of a particular process using potentially dangerous substances may simply mean relocating it in a controlled area, thereby separating the majority of the workforce from the risk. Alternatively, it could involve the construction of a major hazards installation, such as a chemical manufacturing plant in a remote geographical area. In the first case, well-established procedures for limiting access to only trained and authorised operators will be necessary.

Note: Both isolation and enclosure will present maintenance requirements and, in many cases, the operation of a permit to work system will be necessary to safeguard those who undertake this work.

Ventilation engineering applications

Infiltration of air into buildings through openings in the fabric and even planned natural ventilation give no continuing protection wherever toxic gases, fumes, vapours, etc. are emitted from a process. Local exhaust ventilation (LEV) systems must therefore be operated. In certain cases, dilution ventilation may be appropriate.

LEV systems

Broadly, LEV systems are designed to intercept the contaminant as soon as it is generated and direct it into a system of ducting connected to an extractor fan. They ensure that the contaminant is removed from the workplace before it can be inhaled. (See Fig 7.1.)

LEV systems incorporate a number of principal features:

- a hood, enclosure or other inlet to collect and contain the offending agent close to the source of its generation
- ductwork to convey the contaminant away from the source
- a filter or other air-cleaning device to remove the contaminant from the extracted airstream (the filter should normally be located between the hood and the fan)
- a fan or other air-moving device to provide the necessary air flow
- further ductwork to discharge the cleaned air to the outside atmosphere at a suitable point.

HSE booklet HS(G)37 *An Introduction to Local Exhaust Ventilation* provides excellent guidance in the design and specification of LEV systems.

LEV systems range from simple systems serving single machines to extensive plant providing dust control throughout a large factory. The larger systems incorporate complex ductwork runs, which divide into many branches between the air cleaner and the individual machines served by the system (see Fig 7.2). The principles of control at the hood or enclosure are not affected by the complexity of the rest of the system.

To design an effective LEV system the designer will need information on the nature of the contaminant, how it is generated, and the size of the particles produced (if it is a dust or liquid aerosol).

Table 7.1 gives an indication of the physical characteristics of different types of contaminant and Fig 7.3 shows various ways in which contaminant substances can be released into the workroom atmosphere.

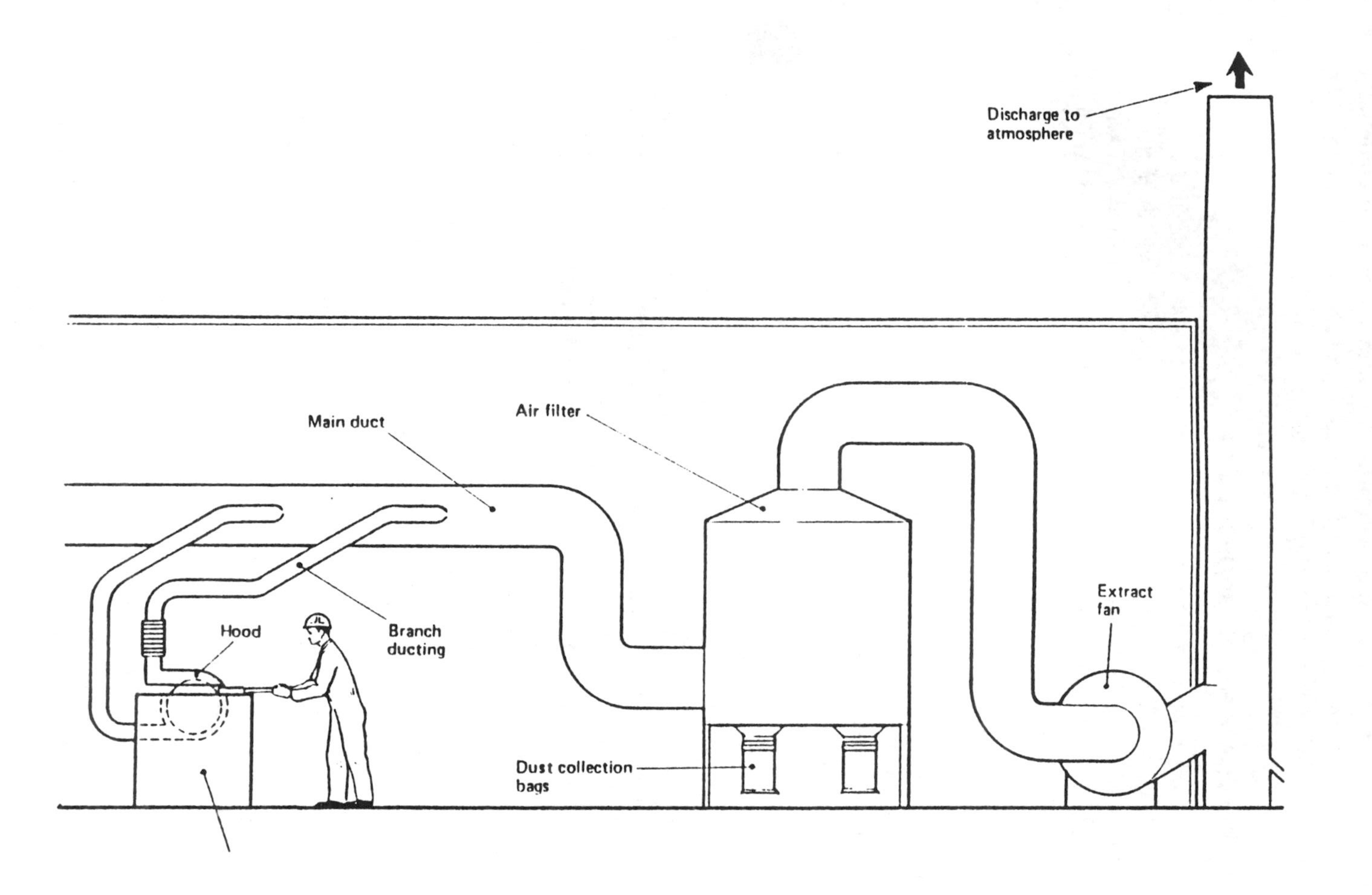

• **FIG 7.1 A typical LEV system**

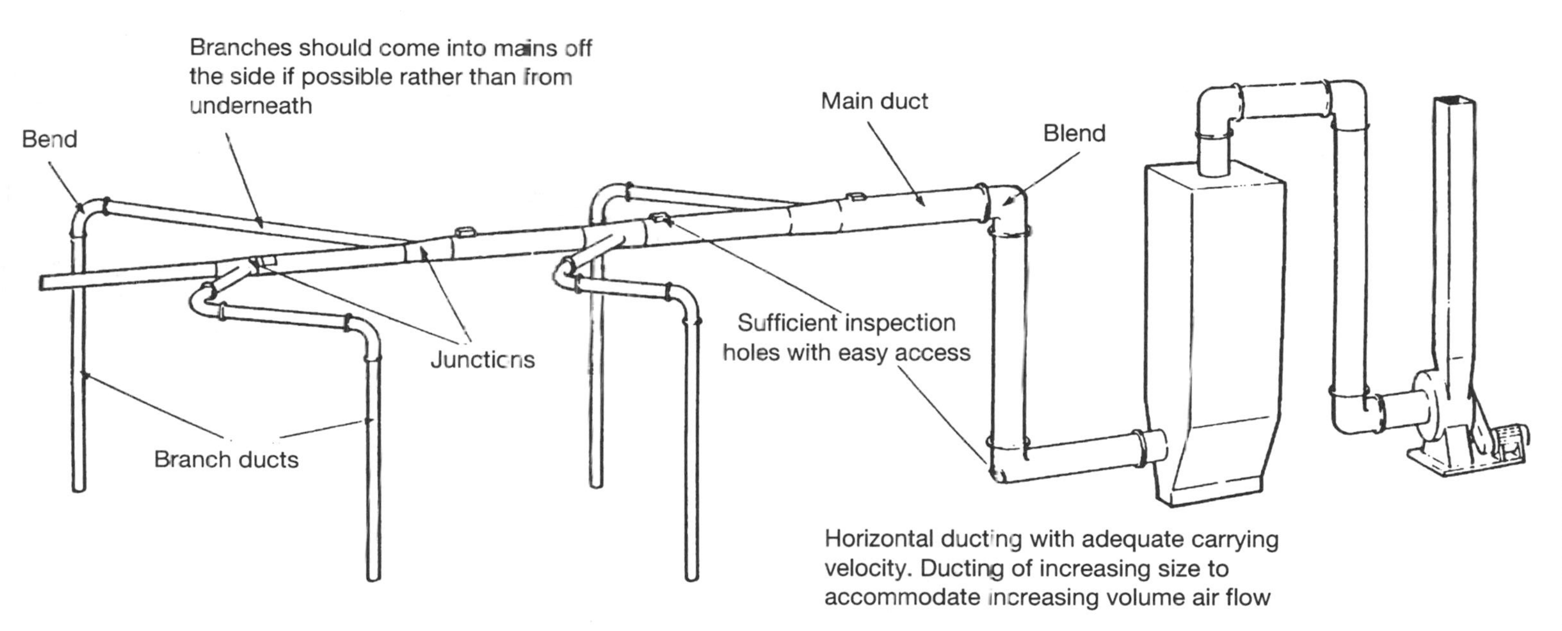

• **FIG 7.2 A more complex LEV system**

TABLE 7.1 Types of airborne contaminant

Contaminant	*Size range* μm	*Characteristics*
Dust	0.1–75	Generated by natural fragmentation or mechanical cutting or crushing of solids (e.g. wood, rock, coal, metal, etc). Grit particles (usually considered to be above 75 μm) are unlikely to remain airborne
Fumes	0.001–1.0	Small solid particles of condensed vapour, especially metals, as in welding or melting processes. Often agglomerate into larger particles as the small particles collide
Smoke	0.01–1.0	Aerosol formed by incomplete combustion or organic matter. Does not include ash e.g. fly ash
Mist	0.01–10.0	Aerosol of droplets, formed by condensation from gaseous state, or as dispersion of liquid state e.g. hot open surface tank, electroplating
Vapour	0.005	Gaseous state of materials that are liquid or solid at normal room temperature and pressure e.g. solvent vapours
Gas	0.0005	Materials which do not normally exist as liquids or solids at normal room temperature and pressure

LEV systems take a number of forms as follows:

Receptor systems

In a receptor system the contaminant enters the system without inducement. The fan in the system is used to provide air flow to transport the contaminant from the hood through ducting to a collection system.

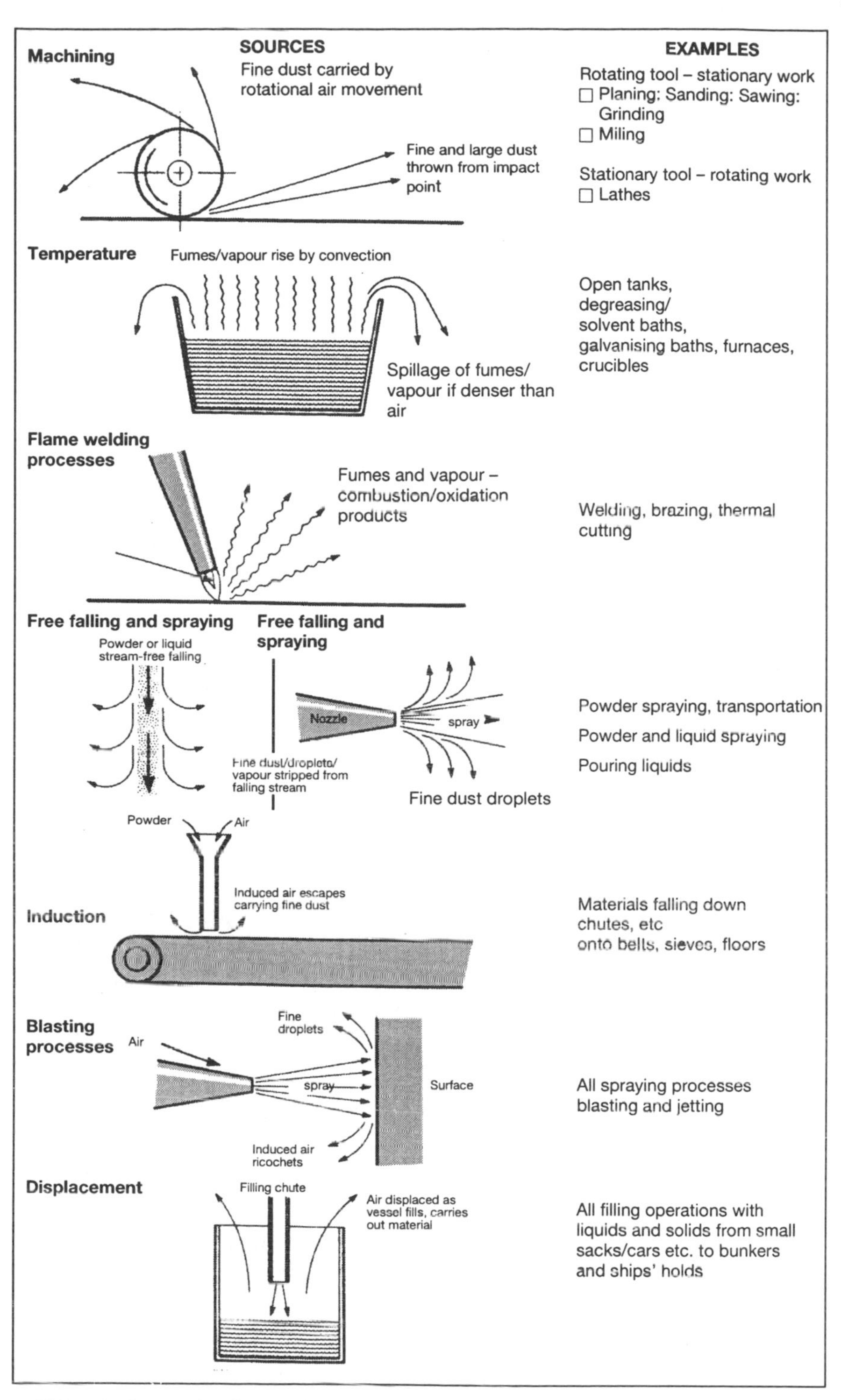

• **FIG 7.3 Sources of contamination**

7

The hood may form almost a **total enclosure** around the source, as with highly toxic contaminants, such as beryllium, and radioactive sources.

In other cases, it may form a **partial enclosure** e.g. a three-sided spray booth in which all spraying takes place. A partial enclosure should be large enough to contain the work and allow it to be manipulated and the process carried out. The air flow within the enclosure should be capable of guiding the contaminant towards the extract point as soon as it is released into the atmosphere. This will require both an adequate air velocity and a booth design which prevents the contaminant spilling out at the front of the enclosure. These points are equally valid for both walk-in spray booths and smaller devices such as fume cupboards and ventilated benches.

In a limited number of situations, the use of a **receptor hood** directly above a source of contaminant may be appropriate in the removal of contaminants from a process. These hoods range in size from small nozzles to large canopies positioned above, below or to the side of the source. They should be located close to the source enclosing it if possible, and designed so that the flow pattern of the air will ensure that the contaminant is captured and retained.

Such hoods, are, however, prone to the effects of side draughts which can lessen the efficiency of the collection system, and their location well away from such draughts is of vital importance.

Generally, hoods which receive the contaminant air as it flows from its origin under the influence of thermal currents are classed as receptors (see Fig 7.4(a)).

Captor systems

With a captor system the moving air captures the contaminant at some point outside the collection hood and induces its flow into it. The rate of air flow into the hood must be sufficient to capture the contaminant at the furthest point of origin, and the air velocity induced at this point must be sufficient to overcome any tendency the contaminant may have to go in any direction other than into the hood. Contaminants emitted with high energy (large particles with high velocity) will require high velocities in the capturing airstream (see Fig 7.4(b)).

Low-volume high-velocity (LVHV) systems

Dust particles emitted by high-speed grinding machines or pneumatic chipping tools require very high capture velocities. One method of achieving these very high velocities at the source is to extract from small apertures very close to the source of the contaminant (see Fig 7.4(c)). Thus high velocities can be achieved with quite low air flow rates.

The following factors are essential in LVHV design:

- appropriate ergonomic design of the cowl and hoses
- correct cowl adjustment
- regular thorough maintenance
- training of operators.

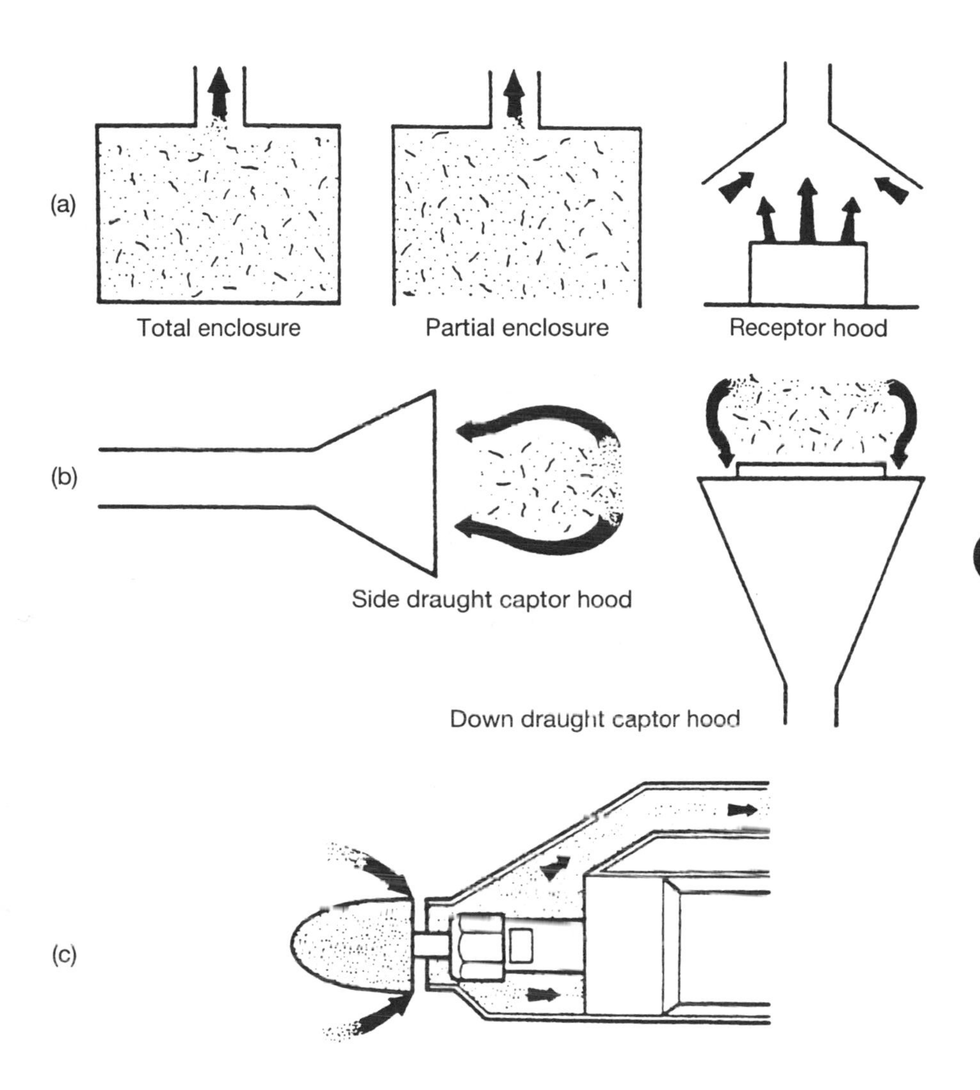

• **FIG 7.4 LEV systems**

Source: American Conference of Government Industrial Hygienists, 1980

(a) Receptor system
(b) Captor system
(c) Low-volume high-velocity system

Further examples of local exhaust ventilation systems are shown in Fig 7.5 to 7.9.

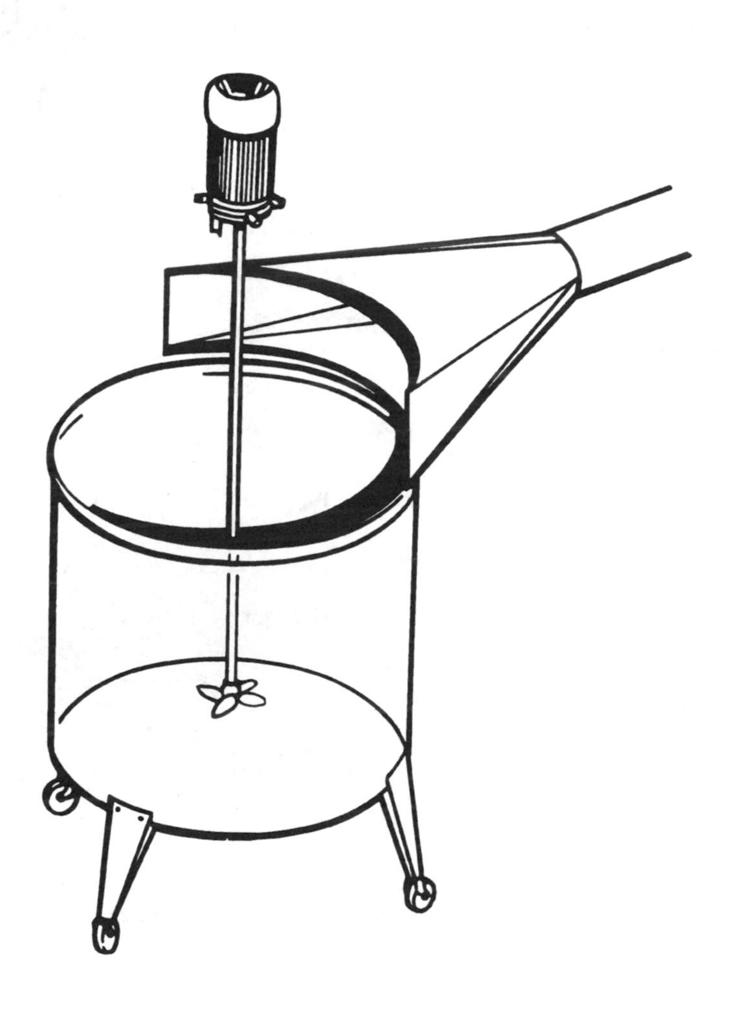

• **FIG 7.5 Local proximity hood applied to paddle mixer**

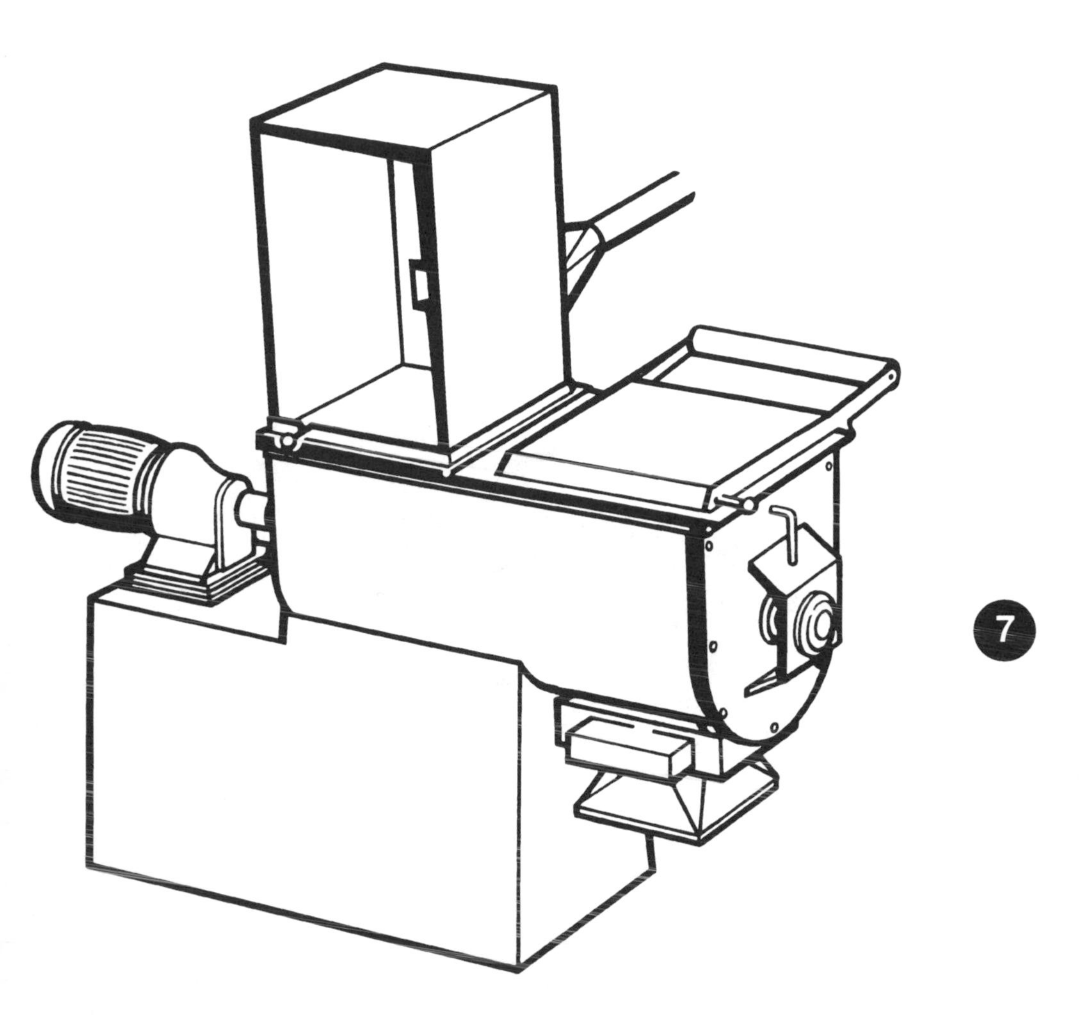

• **FIG 7.6 Enclosure applied to ribbon mixer**

Ventilation system maintenance, examination and monitoring
In addition to effective preventive maintenance by trained persons, there are statutory requirements to undertake formal examination and testing of LEV systems in some industries. For effective maintenance, examination and testing, it is vital to have comprehensive information on the system and design specification i.e. the information which should be provided at commissioning, together with detailed maintenance and test procedures provided by manufacturers and suppliers of the systems.

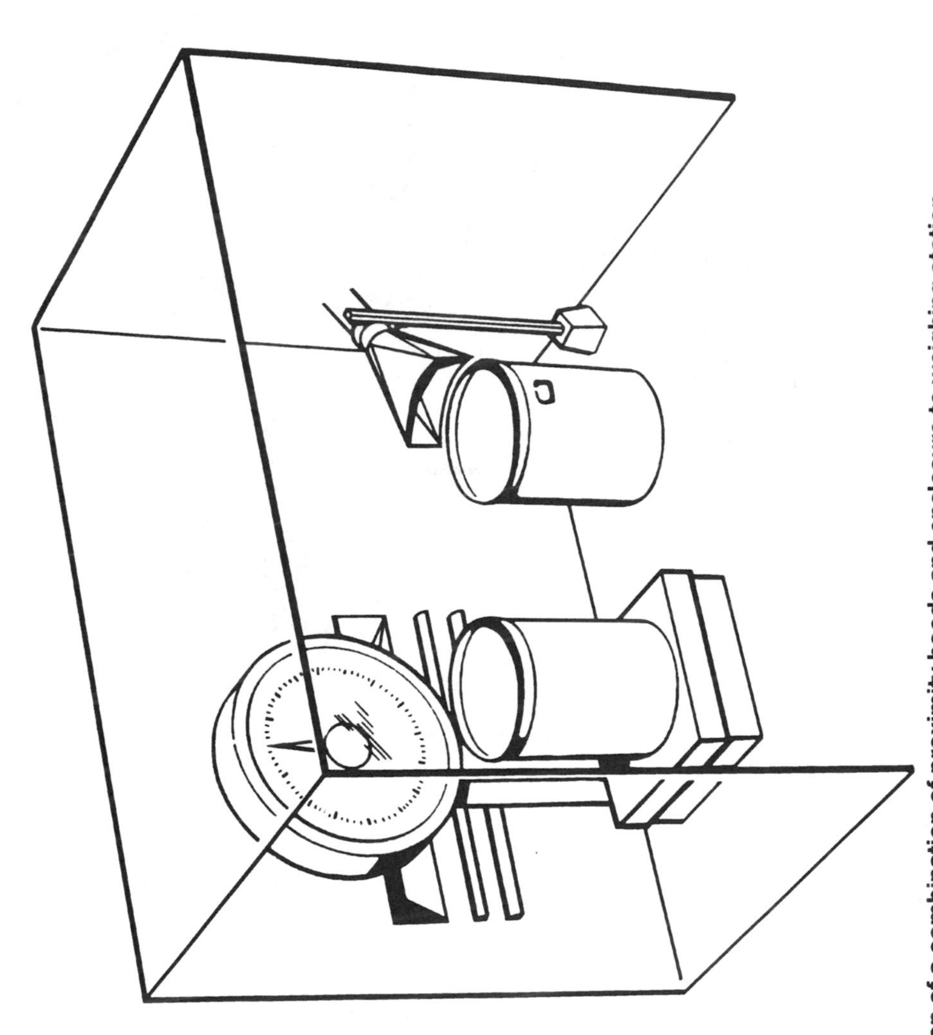

FIG 7.7 Application of a combination of proximity hoods and enclosure to weighing station

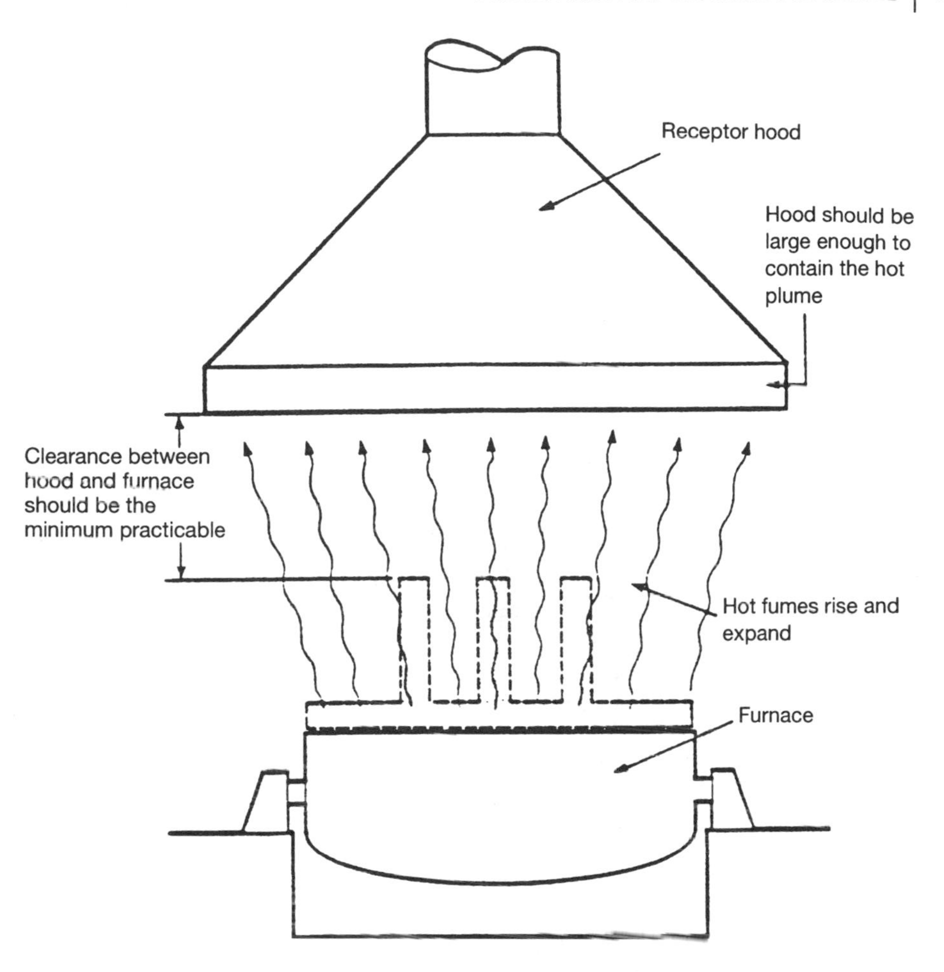

7

• **FIG 7.8 Typical receptor hood**

Comprehensive guidance on these issues is provided in HSE Guidance Note HS(G)54 'The Maintenance, Examination and Testing of Local Exhaust Ventilation'.

Dilution ventilation

In certain situations it may not be possible to extract a contaminant close to the point of origin. If the quantity of contaminant is small, uniformly evolved and of low toxicity, it may be possible to dilute the contaminant by inducing large volumes of air to flow through the contaminated area. Dilution ventilation is most successfully applied to control vapours, such as organic vapours from low-toxicity solvents, but is seldom successfully applied to dust and fumes, as it will not prevent inhalation. In cold weather this method has self-evident implications for cost and thermal discomfort.

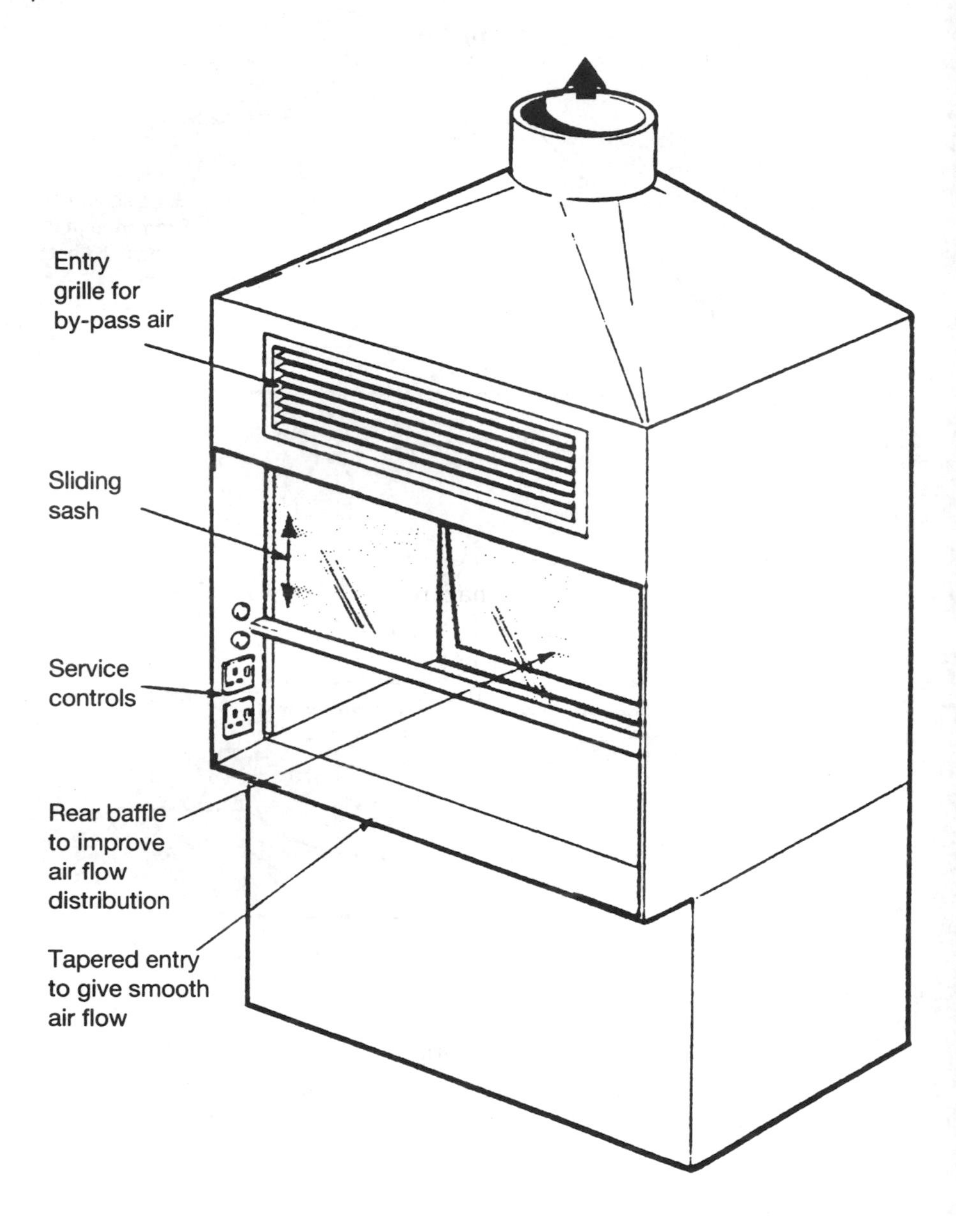

• **FIG 7.9 Laboratory fume cupboard**

Segregation

Segregation is a method of controlling the risks from toxic materials or physical hazards, such as noise and radiation. It can take a number of forms, as follows.

Segregation by distance (separation)

This is the relatively simple process where a person separates himself from the source of the danger. This is appropriate in the case of noise where, as the distance from the noise source increases, the risk of occupational deafness reduces. Similar principles apply in the case of radiation. Segregation by distance protects those at secondary risk, if those at primary risk are protected by other forms of control.

Segregation by age

The need for protection of young workers has reduced over the last 50 years but, where the risk is marginal, it may be necessary to exclude young persons, particularly females, from certain activities. An example of segregation by age as a control strategy occurs in the Control of Lead at Work Regulations 1980, which exclude the employment of young persons in lead processes.

Segregation by time

This refers to the restriction of certain hazardous operations to periods when the number of workers present is small, for instance at night or during weekends, and when the only workers at risk are those involved in the operation. An example of such an operation is the examination by radiation of very large castings.

Segregation by sex

There is always the possibility of sex-linked vulnerability to certain toxic materials, particularly in the case of pregnant women, where there can be damage to the foetus e.g. in processes involving lead.

Change of process

Improved design or process engineering can bring about changes to provide better operator protection. This is appropriate in the case of machinery noise or dusty processes. A typical example is the dressing of seed corn with mercury-based fungicides. Traditionally, this was carried out in a seed dressing machine in which the fungicidal powder was brushed on to the seed corn as it passed along a conveyor inside a dressing chamber. The fungicidal powder escaped through apertures in the structure of the chamber, contaminating the atmosphere and surrounding structural items. A change to liquid seed dressing using a sealed spray chamber has eliminated this toxic dust hazard.

Controlled operation

Controlled operation is closely related to the duty under HSWA section 2(2)(c), to provide a safe system of work. It is particularly appropriate where there is a high degree of foreseeable risk. It implies the need for high standards of supervision and control, and may take the following forms:

- isolation of a process in which dangerous substances are used or where there may be a risk of heat stroke
- the use of mechanical or remote control handling systems e.g. with radioactive substances
- the use of permit to work systems e.g. entry into confined spaces, such as closed vessels, tanks and silos, or fumigation processes using dangerous substances such as methyl bromide
- restriction of certain activities to highly trained and supervised staff e.g. competent persons working in high-voltage switchrooms.

Reduced time exposure (limitation)

Risks to health from dangerous substances or physical phenomena such as noise, can be reduced by limiting the exposure of workers to certain predetermined maxima. This strategy forms the basis for the establishment of occupational exposure limits i.e. long-term exposure limits (8-hour time-weighted average value) and short-term exposure limits (10-minute time-weighted average value). This strategy is encompassed in the Noise at Work Regulations 1989.

Dilution

There is always some danger in handling chemical compounds and wastes in concentrated form. Handling and transport in dilute form tend to reduce the risk. This strategy is appropriate where it is necessary to feed strong chemicals into processing plant regularly or transport quantities of dangerous substances for short distances in open containers. Generally, such practices should be discouraged and, wherever possible, eliminated by a process change. Dilution is a poor form of control strategy.

Neutralisation

This is the process of adding a neutralising compound to another strong chemical compound e.g. acid to alkali, thereby reducing the immediate danger. This strategy is practised commonly in the transportation of strong liquid waste chemical substances e.g. acid-based wastes, where a neutralising compound is added prior to transportation, and in the treatment of industrial effluents prior to their passing into a public sewer.

SUPPORT STRATEGIES

In the successful implementation of the various prevention and control strategies, the use of a number of support strategies may be necessary. These include:

- control over cleaning and housekeeping activities, through the operation of cleaning schedules
- the operation of planned maintenance systems, particularly in the case of LEV systems
- the provision and maintenance of high standards of welfare amenities – sanitation, hand cleansing and shower facilities, and clothing storage arrangements
- strict control over personal hygiene practices
- the selection, provision, use and maintenance of various items of personal protective equipment
- health surveillance of identified groups at risk
- the provision of information, instruction and training
- the use of various forms of health and safety propaganda, such as posters
- effective joint consultation with workers on the risks and precautions necessary.

Maintenance procedures

LEV systems should feature in an organisation's written planned maintenance procedures. Such information should incorporate:

- the frequency of maintenance
- the actual maintenance tasks, including the detection of defects and their correction
- individual responsibility for ensuring the tasks are completed satisfactorily
- any precautions necessary on the part of maintenance staff.

Any maintenance programme should anticipate potential problems and ensure that the LEV plant continues to attain acceptable standards of performance and to control emissions effectively.

Effective maintenance should include:

- regular inspection of the plant including a weekly check for signs of potential damage, wear and/or malfunction
- monitoring of performance indicators e.g. air velocity, static pressures, electrical power consumption
- routine replacement of components known to have a limited useful life e.g. filters

- prompt repair or replacement of components which are found to be worn or damaged.

Procedures for examination and testing

The procedure adopted for the thorough examination and test will depend upon the type of plant and the pollutant it handles. The responsible person will decide on which techniques are to be used based on his knowledge and experience of the plant, published guidance, information from suppliers, appropriate British Standards and on legal requirements. In the latter case, both the Control of Lead at Work Regulations 1980 and the Control of Asbestos at Work Regulations 1987 specify these tests.

The examination and the tests adopted should be sufficient to show that the plant is in good working order, that it meets acceptable performance standards and that emissions are satisfactorily controlled. A thorough examination and test will normally comprise:

- a thorough external and, where appropriate, internal examination of all parts of the system
- an assessment of control, for example, by the use of a dust lamp, static air monitoring and/or smoke testing
- measurement of plant performance, for example, by static pressure measurement behind each hood or enclosure, air velocity measurement at the face of the enclosure or point of emission, air velocity measurement in the duct and/or power consumption
- where air is recirculated, an assessment of the performance and integrity of the air cleaner or filter.

Where possible, the examination and test should be carried out with the process in operation. If the LEV system comprises several booths, hoods or terminals which can be turned on or off by means of control dampers, these should all be open during the test unless there is good reason to believe that the system is never operated in this mode.

Ventilation system monitoring techniques

Techniques can be classified as follows:

- direct measurement of emissions (air monitoring/sampling)
- visualisation techniques
- measurement of plant performance.

Direct measurement

This entails the taking of samples by way of:

- the use of personal dosimeters; and/or
- static sampling techniques.

Visualisation and direct assessment

The **Tyndall beam dust lamp** enables respiratory dust and fumes, which are invisible to the naked eye in normal lighting, to be observed. Although not a quantitative technique, it is probably the most useful method of assessing whether dust and fumes are released. It enables the source of the dust to be identified and its direction and movement to be noted (see Fig 7.10).

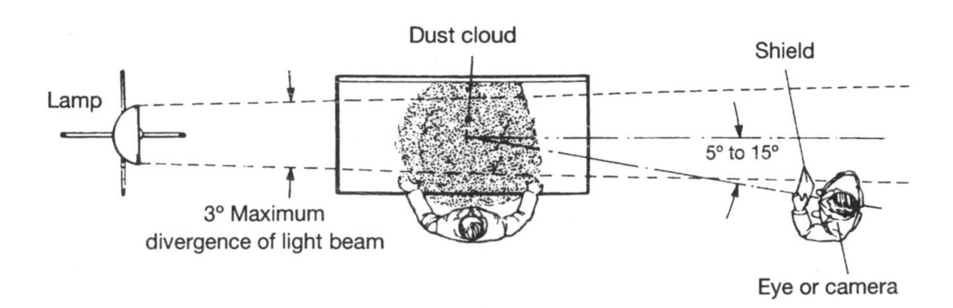

7

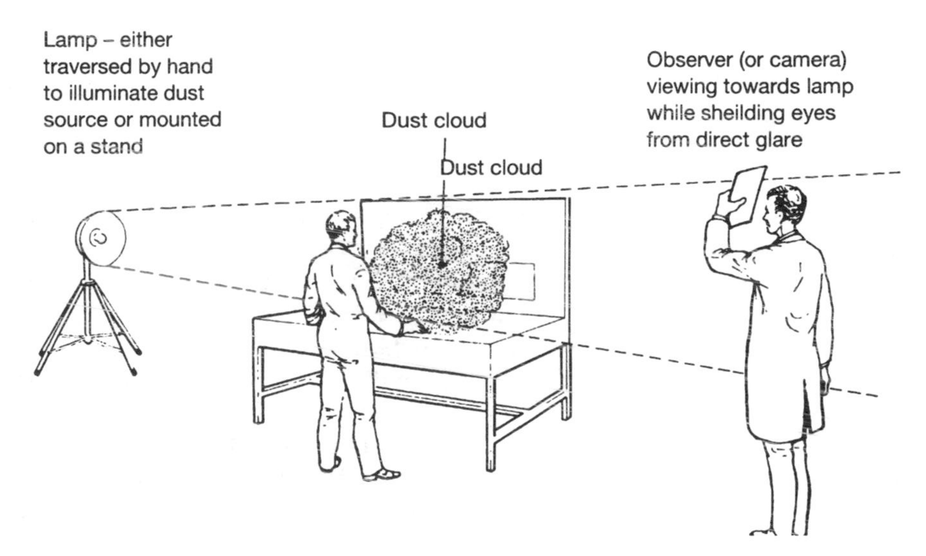

• **FIG 7.10 The use of the dust lamp**

The lamp produces a horizontal beam of light. When the beam passes through a cloud of dust, forward scatter of light occurs and this is visible to an observer looking along the beam of light in the direction of the lamp.

Correct use of the dust lamp is essential and two people may be needed, one to observe the cloud and one to position and hold the lamp, especially if a small portable lamp is used.

Other techniques can be used to visualise the flow of certain pollutants. Such techniques as **schlieren** and **infra-red photography** and **thermography** rely on visualisation of physical properties of the released pollutant, such as density and temperature.

Smoke can be used to detect the direction and pattern of air movement around the source and to show whether air is drawn form the source into the LEV system. When released in sufficient quantity from a smoke generator inside an exhaust ventilated enclosure, it may be possible to detect leakage by looking for signs of escape of smoke.

Containment testing can be used in specific circumstances to test the effectiveness of exhaust ventilated enclosures. A **tracer**, for example a gas, vapour or aerosol is released at a predetermined rate inside the enclosure and then monitored at set points outside the enclosure to determine if any escapes. This is a specialised technique which requires validation for a particular type of enclosure. It does not give an absolute measure of protection but can be compared with a known standard to determine whether performance is adequate.

Tracers can also be used for other purposes, particularly in determining the movement of pollutant from a source where the pollutant itself is difficult to measure or detect and for measuring ventilation rates within a building.

Measurement of LEV plant performance

Assessment of LEV plant performance entails measurement, in particular, of static pressures, air velocity and fan power consumption.

Measurement of static pressure

Static pressure at any point in the system is determined by the air flow rate and the resistance of the system due to friction. Provided there is no change in the air flow system resistance, the static pressure measured (by a manometer) at a particular point will remain constant. If the air flow rate changes, for example, because of fan wear, or if the resistance increases because ductwork is blocked or damaged, then this will usually cause a change in the static pressure at that point (see Fig 7.11).

Ideally, static pressure should be measured in a straight section of duct at a distance equivalent to several times the duct diameter from the hood or any bends, etc. and there should be no damper or means of adjusting air flow between the hood and the test point. In these circumstances it is possible to estimate changes in air flow from the change in measured static pressure.

Instruments for the measurement of static pressure include:

1 *Liquid manometer* A column of liquid of known specific gravity is displaced or drawn up a calibrated column (depending upon whether the static pressure is positive or negative). The simplest liquid manometer is a 'U' tube partially filled with liquid. They give a direct measurement of pressure and are usually supplied with an interchangeable calibrated scale.

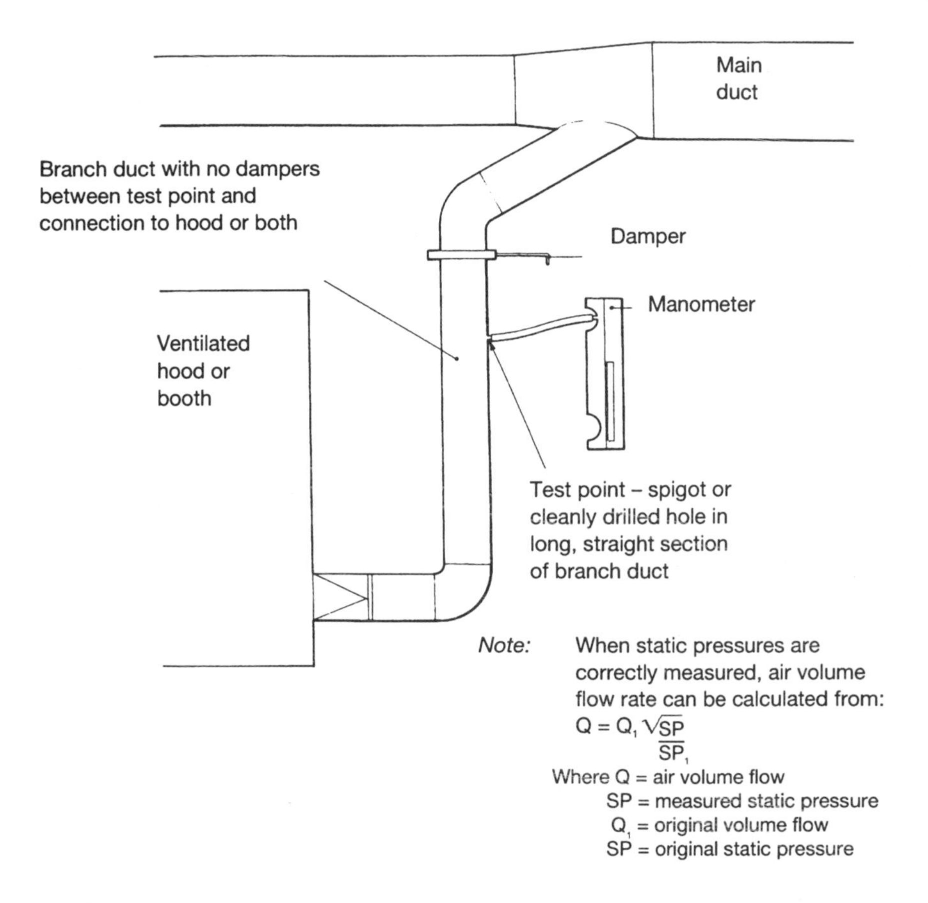

• **FIG 7.11 Correct use of manometer**

2 *Electric manometer* This instrument contains a sealed transducer with a diaphragm or membrane which is deflected as the pressure changes. This deflection is measured electronically and displayed as pressure. They are easy to use but require calibration.
3 *Aneroid pressure gauge* A number of instruments are available in which the change in pressure is converted to mechanical movement such as the Bourdon gauge, the diaphragm gauge and bellows gauge. Aneroid gauges are robust and can operate at very high pressures but they are generally less accurate than manometers.

Measurement of air velocity

Air velocity measurement serves two purposes:

- measurement of air velocities into hoods, etc. or at the working position will indicate whether the airborne pollutant is likely to be adequately controlled (see Table 7.2)

- measurement of duct velocity will enable the examiner to check whether the system is handling the required air volume and whether the velocity is high enough to convey pollutant to the collector. (See Table 7.3)

Air velocity measurement is generally undertaken using various types of anemometer. Commonly used instruments include:

1 *Thermal anemometers* These electronic instruments detect the change of resistance of a heated wire sensor or thermistor when it is cooled by air movement. Response is rapid, probes are small and convenient to use, but instruments need calibrating and are susceptible to damage and fouling. These instruments give no indication of the direction of air flows unless the head is specially shielded.
2 *Rotating vane anemometer* The vane acts as a propeller, the speed of rotating being dictated by air velocity, which can be calculated from the number of rotations of the vane in a set time or displayed automatically. They are not suitable for air velocities below 0.2 m/s and need to be aligned with the air flow so giving a general indication of its direction.
3 *Swinging vane anemometer (Velometer)* This instrument houses a hinged vane which is deflected by the air movement giving a direct scale reading. It is rather cumbersome to use and a range of fittings are required for different applications.
4 *Pitot static tube* This comprises two concentric tubes with openings to measure total and static air pressures from which the air velocity can be calculated. This device can be used to measure air velocities over 3 m/s and needs to be positioned to point into the air flow. They are commonly used for measuring duct velocities and do not need calibration (although the manometer used may be calibrated).
5 *Orifice plate* Air velocity can be calculated from the pressure drop measured across an orifice plate fixed in the duct. This technique is suitable for fixed monitoring for maintenance purposes (see Fig 7.12 (b)).

Air velocity measurements should be accurate and repeatable. The following points should be considered when taking measurements:

- Where the face velocity of a hood or enclosure or the velocity onto a grille or vent is required, a series of measurements should be taken across the plane of the face or opening and averaged. The number of readings will depend upon the size of the opening. Readings which vary more than 20 per cent from the average together with oscillation of the anemometer reading may indicate eddies and reversal of flow (see Fig 7.12 (a)).
- When measuring air velocity in ducts, a series of readings should be taken ideally in a straight section of at least 7.5 duct diameters downstream of any disturbance such as bends, junctions, etc. and the average of these used for calculation of air volume flow. For an approximate estimation of air velocity in circular ducts, the velocity in the centre of the duct may be measured and multiplied by a factor of 0.8. However for an accurate

assessment in circular or rectangular ducts, the number of readings will depend on the size of the duct.

- The anemometer should be carefully chosen so that it is small enough to be completely inserted into the air stream without causing a serious obstruction.
- Readings should, wherever possible, be taken with a hood or booth empty. If this is not practicable, the size and nature of any obstruction should be noted to enable this to be considered when later readings are taken. The person taking the reading should stand back from the opening if possible to minimise disturbance of air flow.

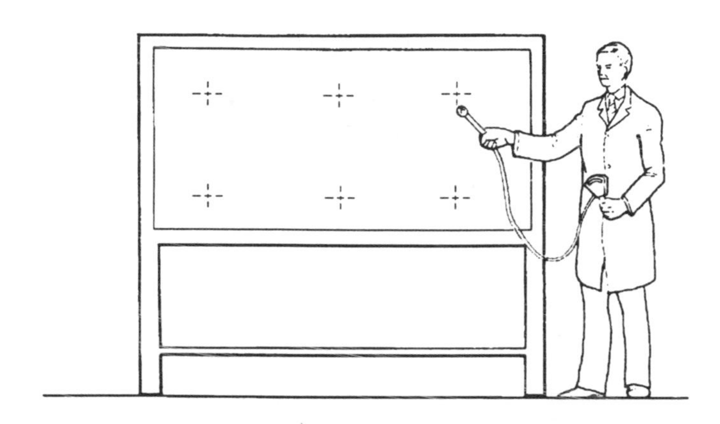

(a) Air velocity is measured at a series of positions across the face of the booth

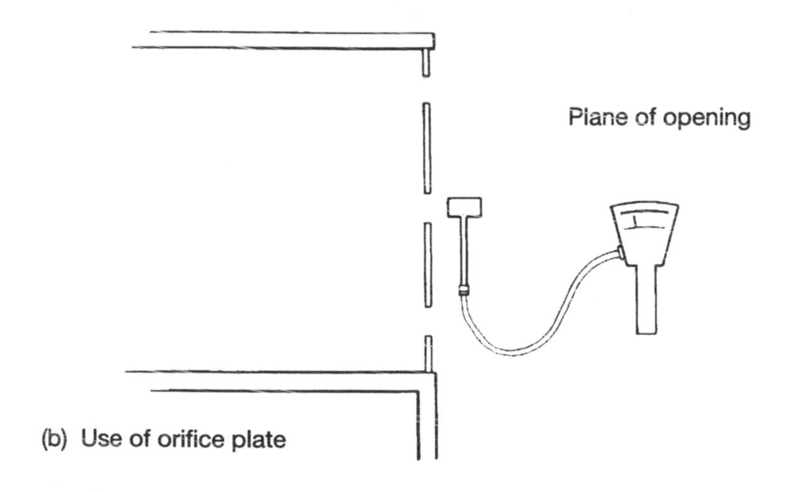

(b) Use of orifice plate

FIG 7.12 Air velocity measurement

7

TABLE 7.2 Range of capture velocities

Conditions of dispersion of contaminant	*Examples*	*Capture velocity:* $m\ s^{-1}$
Released with practically no velocity into quiet air	Evaporation from tanks, degreasing etc.	0.26–0.50
Released at low velocity into moderately still air	Spray booths, intermittent container filling, low-speed conveyor transfers, welding, plating, pickling	0.50–1.00
Active generation into zone of rapid air motion	Spray painting in shallow booths, barrel filling, conveyor loading, crushing	1.00–2.50
Released at high initial velocity into zone of very rapid air motion	Grinding, abrasive blasting, tumbling	2.50–10.00

In each category above, a range of capture velocities is shown. The choice of values depends on several factors:

Lower end or range

(a) Room air currents minimal or favourable to capture
(b) Contaminants of low toxicity or of nuisance value only
(c) Intermittent, low production
(d) Large hood – large air mass in motion

Upper end of range

(a) Disturbing room air currents
(b) Contaminants of high toxicity
(c) High production – heavy use
(d) Small hood – local control only

Adapted from: *Industrial Ventilation* 19th ed, ACGIH

TABLE 7.3 Recommended duct velocities

Type of contaminant	*Duct velocity*: m s^{-1}
Gases (non-condensing)	No minimum limit
Vapours, smoke, fumes	10
Light-medium-density dust (e.g. sawdust, plastic dust)	15
Average industrial dusts (e.g. grinding dust, wood shaving, asbestos, silica)	20
Heavy dusts (e.g. lead, metal turnings, and dusts which are damp or tend to agglomerate)	25

Measurement of fan power consumption

Fan power consumption must be taken into account in any assessment of LEV efficiency. Any deterioration in the LEV plant which leads to a reduction in air flow can be reflected by a reduction in the electrical power consumption of the fan. If the amperage to the fan motor is measured and is below the maker's specification or the figure recorded when the initial appraisal of the LEV system was carried out, then this may indicate that the LEV plant performance has deteriorated and that further investigation is warranted. Monitoring consumption is most appropriate for maintenance purposes.

7

Other measurements may need to be taken if the appointed person is to be satisfied that the plant is in good condition and operates satisfactorily. These include, for example:

- fan and motor speeds
- air temperature if hot gases are handled
- humidity if the process releases steam.

TEMPERATURE, AIR FLOW AND HUMIDITY CONTROL

Comfort

'Comfort' is a subjective assessment of the conditions in which an individual works, sleeps, relaxes, travels, etc. The sensations of comfort vary with a person's age, state of health and vitality. Despite the fact that comfort is a personal state, a degree of unanimity is usually found wherever a group of people are asked to assess a given atmospheric condition. Research indicates that there are four factors which are chiefly responsible for the production of the sensation of thermal comfort, namely air temperature, radiated heat, humidity and the amount of air movement. There are limits to the ability of the human body to adapt to achieve this state of comfort. Few people, for example, would choose to live in a greenhouse or a house without windows all the year round, yet in industry workers may be expected to do just that! The 'indoor climate' is of prime importance. Failure to recognise this fact can result in poor standards of performance and efficiency, discontent, increased labour turnover, increased accident rates and absenteeism.

Temperature

In order to understand why it is necessary to control the thermal environment, the important process of body temperature regulation, or 'thermoregulation', should be considered.

The chemical process for generating heat by food conversion is an important feature of a person's metabolism. Food is a source of energy and the body converts approximately 20 per cent of this energy to mechanical energy, the remaining 80 per cent being utilised as heat.

A healthy person has a body temperature of 36.9°C which is kept remarkably constant, largely by the body continually varying the flow rate of blood. When body temperature increases, blood flows to the skin and dissipates its heat through the skin surface by thermal exchange. Conversely, when body heat is low, heat is conserved in the deep tissues to maintain what is commonly known as the 'core temperature'. If the air temperature is too high and the differential between the skin and the surrounding air is small, insufficient heat is lost through normal exchange. As a result, the body overheats and the sweat glands are activated. In very hot conditions as much as 1 litre of body fluid can be lost each hour. This reduces the body fluid level and salt deficiencies may be created.

Stress conditions

Although reference is made to heat stroke in Chapter 3, it is appropriate here to consider stress conditions associated with extremes of temperature.

Heat stress

Under most industrial working conditions, operators tend to be self-limiting in their thermoregulatory control. The average worker will tend to withdraw from a hot environment or heat source before he becomes liable to heat stroke. Whilst there are no specific heat exposure limits in the UK, threshold limits for permissible heat exposure (**indices of thermal stress**) have been established in the USA. The most commonly used index of thermal stress is that based on physiological observations and related to wet bulb globe temperature shown with a whirling hygrometer.

The following equations are used to calculate the wet bulb globe temperature (WBGT) values:

Outdoor work with solar load:

$$WBGT = 0.7B + 0.2GT + 0.1DB$$

Indoor work, or outdoor work with no solar load:

$$WBGT = 0.7WB + 0.3GT$$

where WB = natural wet bulb temperature
DB = dry bulb temperature
GT = globe thermometer temperature

The natural wet bulb temperature is that recorded from the sling hydrometer without any rotation of the sling. After calculating the WBGT, the number in degrees Celsius is compared with recommended limits of work:rest schedules using either a table or a graph (see Table 7.4 and Fig 7.13).

TABLE 7.4 Maximum permissible wet bulb globe temperature readings

	Workload		
Work:rest schedule (per hour)	*Light*	*Moderate*	*Heavy*
Continuous work	30.0°C	26.7°C	25.0°C
75% work, 25% rest	30.6°C	28.0°C	25.9°C
50% work, 50% rest	31.4°C	29.4°C	27.9°C
25% work, 75% rest	32.2°C	31.1°C	30.0°C

Workers should not be permitted to continue their work when their core temperature reaches 38°C.

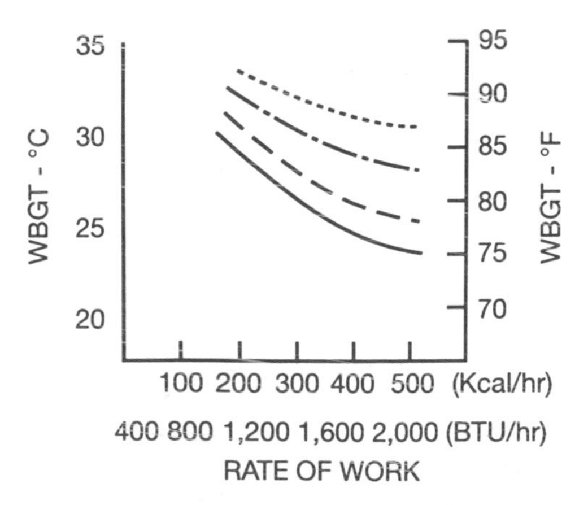

FIG 7.13 WBGT values plotted against rate of work for various work:rest schedules

Cold stress

When outdoor clothing is worn, the limit of tolerance is as shown in Table 7.5. Whilst special protective clothing would be necessary for, say, cold store work, and work at –40°C, some cooling is inevitable. Great attention should be paid in this case to protecting the extremities i.e. hands, feet and head.

TABLE 7.5 Limit of cold stress tolerance

Air temperature	*Time*
–12°C	6 hours
–23°C	4 hours
–40°C	1.5 hours
–57°C	0.5 hours

Ideal comfort conditions

Air temperature

This is the most important factor and depends upon the type of work being undertaken e.g. relatively sedentary work such as office work, light work and heavy (manual) work. For each of these classifications, other thermal conditions notwithstanding, there is either an optimum air temperature or a fairly wide air temperature range (comfort range) within which the majority of workers will not feel any discomfort. It is apparent from the range of activities in Table 7.6 that, as the activity increases, so the optimum air temperature reduces. In any of these categories the air temperature should be reduced when operators are exposed to radiant heat.

TABLE 7.6 Optimum working temperatures

Type of work	*Comfort range*
Sedentary office work Comfort range	 19.4 to 22.8°C
Light work Optimum temperature Comfort range	 18.3°C 15.5 to 20°C
Heavy work Comfort range	 12.8 to 15.6°C

Radiant heat

It is important to regulate the worker's exposure to radiated heat. There should not be a temperature gradient between head and feet of more than 3°C. In fact, any temperature gradient should be negative i.e. cool head, warm feet.

Optimum temperature	18.3°C
Comfort range	16.6 to 20°C

Relative humidity

Relative humidity is defined as 'the actual amount of moisture present in air expressed as percentage of that which would produce saturation'. It is generally accepted that relative humidity should be between 30 and 70 per cent.

If the relative humidity is too low, a feeling of discomfort is produced due to drying of the throat and nasal passages. Conversely, high relative humidity produces a feeling of stuffiness and reduces the rate at which body moisture evaporates (sweat), thereby reducing the efficiency of the body's thermoregulatory system.

Air flow

7

Air flow is an important factor in the consideration of comfort conditions. Movement of air is just perceptible at about 9 metres per minute and complaints of draughts will be received when it exceeds 30 metres per minute. Below 6 metres per minute, a room could be considered airless. The sensation of air flow is directly related to air temperature and skin sensitivity. If the air is cool, even slight draughts are detectable, whereas if the air temperature is controlled, draughts, although still present, may not be detected. Air flow assists in cooling and causes distress if excessive.

LIGHTING

Two specific health and safety-related aspects are relevant in relation to lighting at work:

- a gradual deterioration in an individual's visual acuity and performance
- the basis for lighting design and specific applications, as in the case of display screen equipment (visual display units).

In any consideration of lighting, including lighting deficiencies and their control, it is important to distinguish between the quantitative and qualitative aspects of lighting.

Quantitative aspects

The quantity of light flowing from a source such as a light bulb or fluorescent light (luminaire) is the luminous flux or light flow, which is generally termed 'illuminance'. The units of measurement of luminous flux were formerly foot candles or lumens per square foot, but more recently this unit has become the lux, which is the metric unit of measurement. Thus:

Foot candles = lumens per square foot
Lux = lumens per square metre
10.76 lux =1 lumen per square metre
1 lux = 0.093 lumens per square metre

On this basis, a conversion factor of 10 or 11 is used for converting from lumens per square foot to lux i.e. 20 lumens per square foot = 200 lux.

The illuminance value in lux is measured using a standard photometer or light meter and is the quantity of light present at a particular point.

TABLE 7.7 Average illuminances and minimum measured illuminances for different types of work

General activity	*Typical locations/ type of work*	*Average illuminance* (lux)	*Minimum measured Illuminance* (lux)
Movement of people, machines and vehicles[1]	Lorry parks, corridors, circulation routes	20	5
Movement of people, machines and vehicles in hazardous areas; rough work not requiring any perception of detail[1]	Construction site clearance, excavation and soil work, docks, loading bays, bottling and canning plants	50	20
Work requiring limited perception of detail[2]	Kitchens, factories, assembling large components, potteries	100	50
Work requiring perception of detail[2]	Office, sheet metal work, bookbinding	200	100
Work requiring fine perception of detail[2]	Drawing offices, factories assembling electronic components, textile production	500	200

Note:

(1) Only safety has been considered, because no perception of detail is needed and visual fatigue is unlikely. However, where it is necessary to see detail to recognise a hazard or where error in performing the task could put someone else at risk, for safety purposes as well as to avoid visual fatigue, the figure should be increased to that for work requiring perception of detail.

(2) The purpose is to avoid visual fatigue: the illuminances will be adequate for safety purposes.

Lighting standards are detailed in HSE Guidance Notes HS(G)38 'Lighting at Work'. The Guidance Note distinguishes between **average illuminances** and **minimum measured illuminances** according to the general activity undertaken and the type of work undertaken and typical work location (see Table 7.7).

The ratio between working areas and adjacent areas is also featured in the Guidance Note (see Table 7.8). The Guidance Note recommends that where there is conflict between the recommended average illuminance shown in Table 7.7 and the maximum ratios of illuminance in Table 7.8, the higher value should be taken as the appropriate average illuminance.

TABLE 7.8 Maximum ratios of illuminance for adjacent areas

Situations to which recommendation applies	*Typical location*	*Maximum ratio of illuminances Working area : Adjacent area*
Where each task is individually lit and the area around the task is lit to a lower illuminance	Local lighting in an office	5:1
Where two working areas are adjacent, but one is lit to lower illuminance than the other	Localised lighting in a works store	5:1
Where two working areas are lit to different illuminances by a barrier, but there is frequent movement between them	A storage area inside a factory and a loading bay outside	10:1

Qualitative aspects

The concept of average illuminances in the Guidance Note relates only to the quantity of light, and in the design and assessment of lighting installations consideration must be given to the qualitative aspects.

Factors which contribute to the quality of lighting include the presence or absence of glare in its various forms, the degree of brightness, the distribution of light, diffusion, colour rendition, contrast effects and the system for lighting maintenance.

Glare

This is the effect of light which causes discomfort or impaired vision, and is experienced when parts of the visual field are excessively bright compared with the general surroundings. This usually occurs when the light source is directly in line with the visual task or when light is reflected off a given surface or object. Glare is experienced in three different forms:

1 *Disability glare* is the visually disabling effect caused by bright bare lamps directly in the line of sight. The resulting impaired vision (dazzle) may be hazardous if experienced when working in high-risk processes, for example at heights or when driving. It is seldom experienced in workplaces because most bright lamps e.g. filament and mercury vapour, are usually partly surrounded by some form of fitting.
2 *Discomfort glare* is caused mainly by too much contrast of brightness between an object and its background, and is associated with poor lighting design. It causes visual discomfort without necessarily impairing the ability to see detail, but over a period can cause eye strain, headaches and fatigue. Discomfort glare can be reduced by:

   - careful design of shades which screen the lamp
   - keeping luminaires as high as practicable
   - maintaining luminaires parallel to the main direction of lighting.

3 *Reflected glare* is the reflection of bright light sources on shiny or wet work surfaces, such as glass or plated metal, which can almost entirely conceal the detail in or behind the object which is glinting. Care is necessary in the use of light sources of low brightness and in the arrangement of the geometry of the installation, so that there is no glint at the particular viewing position.

Distribution

The distribution of light, or the way in which light is spread, is important in lighting design. Poor lighting design may result in the formation of shadowed areas which can create dangerous situations particularly at night. For good general lighting, regularly spaced luminaires are used to give evenly distributed illuminance. This evenness of illuminance depends upon the ratio between the height of the luminaire above the working position and the spacing of the fittings.

Colour rendition

This term refers to the appearance of an object under a given light source compared to its colour under a reference illuminant e.g. natural light. Colour rendition enables the colour appearance to be correctly perceived. The colour rendering properties of light fitments should not clash with those of natural light, and should be equally effective at night when there is no daylight contribution to the total illumination of the workplace.

Brightness

Brightness, or more correctly, 'luminosity', is essentially a subjective sensation and cannot be measured. It is possible, however, to consider a brightness ratio, which is the ratio of the apparent luminosity between a task object and its surroundings. To achieve the recommended brightness ratio, the reflectance of all surfaces in the workplace should be carefully maintained and consideration given to reflectance value in the design of interiors. Given a task illuminance factor of 1, the effective reflectance values should be:

Ceilings	–	0.6
Walls	–	0.3 to 0.8
Floors	–	0.2 to 0.3

Diffusion

This is the projection of light in many directions with no directional predominance. The directional effects of light are just as important as the quantity of light, however, as the directional flow of light can often determine the density of shadows, which may affect safety. Diffused lighting can soften the output from a particular source and so limit the amount of glare that may be encountered from bare fittings.

Stroboscopic effect

All lamps that operate from an alternating current electricity supply produce oscillations in light output. When the magnitude of the oscillations is great and their frequency is a multiple or sub-multiple of the frequency of movement of machinery, that machinery will appear to be stationary or moving in a different manner. This is called the 'stroboscopic effect'. It is not common with modern lighting systems but where it does occur it can be dangerous, so appropriate action should be taken to avoid it. Possible remedial measures include:

- supplying adjacent rows of light fittings from different phases of the electricity supply
- providing a high-frequency supply
- washing out the effect with local lighting which has much less variation in light output e.g. tungsten lamp
- use of high-frequency control gear if applicable.

Lighting maintenance

A well-organised lighting programme is necessary for permanently good illumination to be achieved. This programme should incorporate regular cleaning and replacement of lamp fittings as a basic consideration, together with regular assessment of illuminance levels with a standard photometer at predetermined point. Furthermore, the actual function of the lighting provided should be reviewed in line with changes that may be made in production, storage or office arrangements. To facilitate safe lamp cleaning and replacement, high-level luminaires should be fitted with raising and lowering gear, so that this work can be undertaken at floor level.

8

Personal protective equipment

Personal protection implies the provision and use of various types of personal protective equipment (PPE), such as safety boots, ear protectors, aprons and gloves. As such, it must be considered as either the last resort, when all other methods of protection have failed, or purely as an interim form of protection until the hazard can be eliminated at source or controlled by some form of 'safe place' strategy, such as machinery guarding or the installation and use of local exhaust ventilation systems.

Personal protective equipment includes a wide range of equipment worn and used by people at work to protect them from both general and specific hazards. As such, it includes:

- *Head protection* safety helmets, bump caps, caps and hair nets.
- *Eye protection* goggles, safety spectacles, visors, hand-held or freestanding screens.
- *Face protection* face shields which can be hand-held, fixed to a helmet or strapped to the head.
- *Respiratory protection* general-purpose dust respirators, positive pressure-powered dust respirators, helmet-contained positive pressure respirators, gas respirators, emergency escape respirators, airline breathing apparatus, self-contained breathing apparatus.
- *Hearing protection* ear plugs, ear defenders, muffs and pads, ear valves, acoustic wool.
- *Skin protection* barrier creams.
- *Body protection* one-piece and two-piece overalls, donkey jackets, aprons, warehouse coats, body warmers, oilskin overclothing.
- *Hand and arm protection* general-purpose fibre gloves, PVC fabric gauntlets, gloves and sleeves, chain mail hand and arm protectors.
- *Leg and foot protection* safety boots, shoes and wellingtons, gaiters and anklets.

FORMS OF PPE

Head protection

There are four widely used types of head protection:

- crash helmets, cycling helmets, riding helmets and climbing helmets
- industrial safety helmets which can protect against falling objects or impact with fixed objects
- industrial scalp protectors (bump caps), which can protect against striking fixed obstacles, scalping or entanglement
- caps, hairnets, etc. which can protect against scalping or entanglement.

Much will depend upon the type and form of the risk an individual is exposed to in terms of the head protection required. Certain activities involving building work are subject to the Construction (Head Protection) Regulations 1989. For head protection to be suitable, it must fit the wearer properly, be of an appropriate size and have an easily adjustable headband, nape and chin strap. The range of size adjustments should be large enough to accommodate thermal liners used in cold weather.

Head protection should not hinder the work being done, and should be compatible with other PPE, such as ear and eye protectors, being worn at the same time.

To ensure maintenance in a good condition, head protection should be stored, when not in use, in a safe place, be visually inspected regularly for signs of damage, have defective harness components replaced and have the sweat band regularly cleaned or replaced.

Eye and face protection

Eye protection serves to guard against the hazards of impact, splashes from chemicals or molten metal, liquid droplets (chemical mists and sprays), dusts, gases, welding arcs, non-ionising radiation and the light from lasers.

There are four principal forms of eye protection:

- safety spectacles, which may incorporate optional sideshields, with lenses manufactured in tough optical plastic, such as polycarbonate, and available with standard or prescription lenses
- eyeshields designed with a frameless one-piece moulded lens
- safety goggles, manufactured with toughened glass lenses or wide vision plastic lenses, with a flexible plastic frame and elastic headband
- faceshields, which are fitted with an adjustable head harness, and provide protection to the face as well as the eyes.

The lenses of eye protectors must be kept clean as dirty lenses restrict vision, can cause visual fatigue and may be a contributory feature in accidents. Eye protectors in a suitable protective case should be issued on a personal basis, and lenses that become scratched or pitted should be replaced.

Respiratory protection

The use of respiratory protective equipment (RPE) is essential wherever workers are exposed to dangerous concentrations of toxic or fibrogenic dusts, fumes or where they may be working in unventilated or poorly ventilated areas. The correct selection and use of RPE is absolutely vital. For instance, the use of face masks is definitely not recommended as a means of protection against anything other than low concentrations of nuisance particulates and atomised liquids. Reference should be made to BS 4275, 'Recommendations for the Selection, Use and Maintenance of Respiratory Protective Equipment' and to the quoted nominal protection factor for the equipment.

Nominal protection factor

BS 4275 refers to the selection of RPE and lists nominal protection factors for different forms of equipment. The nominal protection factor (NPF) measures the theoretical capability of RPE and is calculated thus:

$$\text{NPF} = \frac{\text{Concentration of contaminant in the atmosphere}}{\text{Concentration of contaminant in the face-piece}}$$

Forms of RPE

- **Face masks** These are a simple device for holding filtering media against the nose and mouth to remove coarse nuisance dust particles or non-toxic paint sprays. They should not be used as a means of protection against hazardous or toxic substances.
- **General-purpose dust respirators** These take the form of an ori-nasal face mask and a particulate filter to trap finely divided solids or liquid particles.
- **Positive pressure-powered dust respirators** These comprise an ori-nasal face mask fitted to a power-driven pack carried on the individual and connected by a flexible hose. They are more effective than the simple form of dust respirator as they utilise a much more efficient filtering medium and operate with positive pressure in the face-piece.
- **Helmet-contained positive pressure respirators** This form of device provides head, eye, face and lung protection together with a high degree of comfort. It incorporates a helmet and visor with a high-efficiency axial fan mounted at the rear of the helmet, which draws the dust-laden air through a coarse filter. The partially filtered air is then passed through a fine filter bag. The filtered air provides a cool pleasant air stream over the entire facial area, and is finally exhausted at the bottom of the visor at a flow rate sufficient to prevent dust entering the mouth or nose. The low-voltage

electrical power is supplied by a lightweight rechargeable battery pack, connected to the helmet by means of a flexible cable. This portable battery pack may be clipped to a belt or carried in an overall pocket.

- **Gas respirators** This respirator takes two forms, cartridge and canister. The **cartridge respirator** is similar to the dust respirator. It uses a chemical cartridge filter and is effective against relatively low concentrations of non-toxic gases or vapours which have an acceptable level of concentration exceeding 100 ppm. **Canister respirators**, on the other hand, are normally of the full face-piece type, with exhalation valves, incorporating goggles and visor. They are connected to a chemical canister filter for protection against low concentrations of designated toxic gases or vapours.

  The manufacturer's instructions on the avoidance of cartridge/canister saturation, maximum periods of use in relation to concentrations of gases in air, shelf-life, etc. must, in all cases, be carefully followed. They are effective against toxic gases and vapours in limited concentrations. A particulate filter can be incorporated to remove dust particles.
- **Emergency escape respirators** These are specially designed respirators using a chemical filter which will enable people to escape from dangerous atmospheres in an emergency. They are intended for very short-term use and should never be used for normal protection.
- **Airline breathing apparatus** This apparatus consists of a full mask or half mask connected by flexible hose either to a source of uncontaminated air (short distance) or to a compressed airline via a filter and demand valve. The apparatus is usually safe for use in any contaminated atmosphere (see the manufacturer's stated nominal protection factor) but is limited by the length of the airline, which also places some restriction on movement. When using a fresh air hose, a pump is necessary for lengths over 10 metres. Reference should be made to BS4667, 'Breathing Apparatus'.
- **Self-contained breathing apparatus** This device can be of the open or closed circuit type. The **open circuit type** supplies air by a lung-governed demand valve or pressure reducer connected to a full face-piece via a hose supply. The hose is connected to its own compressed air or oxygen supply which is carried by the wearer in a harness. The **closed circuit type** incorporates a purifier to absorb exhaled carbon dioxide. The purified air is fed back to the respirator after mixing with pure oxygen.

  Both types of device may be used in dangerous atmospheres or where there is a deficiency of oxygen, or for rescue purposes from confined spaces. BS 4667 is relevant in the case of self-contained breathing apparatus.

8

Hearing protection

Under the Noise at Work Regulations 1989 an employer must make hearing protection available to employees where they are likely to be exposed to the first action level (85 dBA) or above. Where employees are exposed to the second action level (90 dBA) or above, or to the peak action level or above, employers must ensure that employees are provided with hearing protection which, when properly worn, can reasonably be expected to keep the risk of damage to those employees' hearing to below that arising from exposure to the second action level or, as the case may be, the peak action level.

When selecting hearing protection, therefore, it is important to ensure that the form of hearing protection – acoustic wool, ear plugs, ear muffs – will produce the necessary attenuation (sound pressure reduction) at the operator's ear.

Ear plugs

These are manufactured in plastic, rubber, glass down or a combination of these materials and are fitted into the auditory canal. They may be of the permanent or disposable type. Training of potentially exposed people in the correct method of inserting the ear plug into the auditory canal is essential. Ear plugs can tend to move out of position with jaw movements.

Ear defenders, muffs and pads

These cover the whole ear and can reduce exposure by up to 50 dBA at certain frequencies. They can be uncomfortable in hot conditions and may be difficult to wear with safety spectacles or goggles. Many safety helmets incorporate a fitting for attaching ear defenders.

Ear valves

These are inserted into the auditory canal and, in theory, allow ordinary conversation to take place while preventing harmful noise reaching the ear.

Skin protection

Occupational dermatitis is the most common form of occupational disease. In many cases it may be associated with poor levels of personal hygiene and the failure to wash chemical and other contamination from exposed skin.

There are many chemical substances which cause dermatitis and are commonly used in the workplace. This includes strong acids and alkalis, chromates and bichromates, formaldehyde, organic solvents, resins, certain adhesives, suds, degreasing compounds and lubricants. Paraffin and trichloroethylene, for instance, remove the natural fats from the skin and render it vulnerable to damage from other substances.

Where dermatitis is identified, medical aid should be sought. In many cases, the presence of dermatitis among operators is a first indication of exposure to hazardous substances, in particular primary irritants and secondary cutaneous sensitisers. As an additional precaution, therefore, the use of barrier creams is recommended. The main objective must be that of preventing exposure to these substances in the first place.

A range of barrier creams is available to meet varying work conditions.

They provide skin protection in wet conditions, and for workers handling acids, alkalis and other potentially hazardous substances. The barrier cream must be applied to the hands and forearms before commencing work.

Body protection

A wide range of protective clothing for the body is available, including:

- coveralls, aprons and overalls to protect against chemicals and other hazardous substances
- outfits to protect against cold, heat and bad weather
- specific clothing to protect against machinery, such as chain saws
- high-visibility clothing
- life jackets and buoyancy aids.

In the selection of body protection, the following factors are relevant:

- the degree of personal contamination from the task or process e.g. dust, oil, general soiling, chemicals, greases, etc.
- the level of hygiene control necessary for a particular product or service e.g. food manufacture, catering
- whether a wet or dry process is involved
- the ease and cost of washing or dry cleaning
- individual preferences shown for specific types of body protection
- the degree of exposure to temperature and humidity variations
- possible discomfort produced by moisture/perspiration when wearing individual garments
- the ease of storage
- the extent to which body protection may restrict movement
- the need for early recognition in hazardous situations e.g. traffic accidents, as with high-visibility clothing
- the potential for life-threatening situations e.g. drowning, as with buoyancy aids and life jackets.

Hand and arm protection

Damage to the hands and arms can arise through the use of machinery and hand tools and in manual handling operations, resulting in cuts and abrasions, through skin irritation, contact with hazardous substances and as a result of exposure to adverse weather conditions.

A wide variety of hand and arm protection is available including gloves and gauntlets made from leather, chain mail, PVC fabric and man-made fibres. Gloves or other hand protection should be capable of giving protection from identified hazards and fit the user. As such, they should be maintained in a good condition, checked regularly and replaced if worn or damaged. Gloves contaminated by chemicals should be washed as soon as possible and before their removal from the hands. Internal contamination of gloves by chemicals is particularly dangerous due to the risk of such substances penetrating the skin and, in these cases, gloves should be discarded.

Leg and foot protection

Safety footwear is used in many industries and occupations e.g. construction, mechanical and manual handling activities, for work in cold and wet conditions, foundry work and forestry. As with other items of PPE, correct selection of leg and foot protection is essential in protecting operators from a wide range of injuries.

Safety boots and shoes

These are the most common type of safety footwear and commonly incorporate a steel toe cap. They may also have other safety features, such as slip-resistant soles, steel insoles and insulation against extremes of temperature.

Clogs

Wooden clogs, frequently fitted with steel toe caps, are traditionally used in a number of industries.

Foundry boots

These incorporate steel toe caps, are heat resistant and designed to protect the foot against molten metal splashes and spillages. They are designed without external features such as laces which trap molten metal and commonly have elasticated sides for quick removal.

Wellington boots

These protect against water and wet conditions and are useful in occupations where footwear needs to be washed and disinfected for hygienic reasons, such as with slaughtermen, agricultural workers and food industry workers. They are manufactured in rubber, polyurethane and PVC, and may feature corrosion-resistant steel toe caps, rot-proof insoles, steel midsoles, ankle bone padding and cotton linings.

Gaiters

These are commonly used in foundries to provide protection to the ankles from splashes of molten metal. They are manufactured in leather and compositions of leather, hessian and other fibrous materials.

Anti-static footwear

These prevent the build-up of static electricity on the wearer. They reduce the danger of igniting a flammable atmosphere and give some protection against electric shock.

Conductive footwear
This type of footwear also prevents the build-up of static electricity, and is particularly suitable for handling sensitive components or substances e.g. explosive detonators. It gives no protection against electric shock.

Leg and foot protection should be subject to regular examination, including the removal of materials lodged in the tread. Waterproofing with silicone-based sprays is recommended to give extra protection against wet conditions.

LIMITATIONS IN THE USE OF PPE

The use of any form of PPE should, in the majority of cases, be seen either as:

- an interim measure until an appropriate 'safe place' strategy e.g. machine guarding, can be implemented; or
- the last resort, when all other protection strategies have failed.

It is never the perfect solution to protecting people from hazards due to the need for users to wear or use the equipment *all the time* they are exposed to such hazards. People simply do not do this for a number of reasons. For instance:

8

- it may create discomfort, restrict movement and be difficult to put on or remove
- it may obscure vision
- it may reduce people's perception of hazards
- it may be inappropriate to the risk e.g. respiratory protection
- it requires, in many cases, frequent cleaning, replacement of parts, maintenance or some form of regular attention by the user, which the operator may see as a chore
- some people perceive the use of personal protective equipment as unnecessary, a sign of immaturity or yet another management imposition; on the whole there is a general reluctance to wear or use same.

SELECTION OF PPE – THE KEY ISSUES

A systematic approach is essential to ensure that workers at risk are protected properly.

When considering the type and form of equipment to be provided, the following factors are relevant:

- the needs of the user in terms of comfort, ease of movement, convenience in putting on, use and removal, and individual suitability

- the number of personnel exposed to a particular hazard e.g. noise
- the type of hazard e.g. fumes, dust, molten metal splashes
- the scale of the hazard
- standards representing recognised 'safe limits' for the hazard e.g. HSE Guidance Notes, British Standards
- specific regulations currently in force e.g. Noise at Work Regulations 1989
- specific job requirements or restrictions e.g. work in confined spaces, roof work
- the presence of environmental stressors which will affect the individual wearing or using the equipment e.g. extremes of temperature, inadequate lighting and ventilation, background noise
- ease of cleaning, sanitisation, maintenance and replacement of equipment and/or its component parts.

Official guidance on the selection, use and maintenance of personal protective equipment is provided in Part 2 of the HSE Guidance to the Personal Protective Equipment at Work Regulations 1992.

The legal requirements

The law surrounding occupational health and hygiene is incorporated in a number of general duties under the HSWA and in various regulations made under this Act, such as the Control of Substances Hazardous to Health (COSHH) Regulations 1994.

This chapter provides a general overview of these principal legal requirements. Detailed guidance is provided in 'Guide to Health and Safety in Practice: Health and Safety Law', including the varying levels of duties, classified as follows:

(a) those of an absolute or strict nature, qualified by the term 'shall' or 'must';
(b) those of a 'practicable' nature; and
(c) those of a 'reasonably practicable' nature.

HEALTH AND SAFETY AT WORK etc. ACT 1974

Section 2 of HSWA places a general duty on employers, so far as is reasonably practicable, to ensure the **health**, safety and welfare at work of all their employees. More particularly, this includes:

(a) the provision and maintenance of plant and systems of work that are, so far as is reasonably practicable, safe and **without risks to health**;
(b) arrangements for ensuring, so far as is reasonably practicable, safety and **absence of risks to health** in connection with the use, handling, storage and transport of articles and substances;
(c) the provision of such information, instruction, training and supervision as is necessary to ensure, so far as is reasonably practicable, the **health** and safety at work of employees;
(d) so far as is reasonably practicable, as regards any place of work under the employer's control, the maintenance of it in a condition that is safe and **without risks to health** and the provision and maintenance of means of access to and egress from it that are safe and without such risk; and

(e) the provision and maintenance of a working environment for employees that is, so far as is reasonably practicable, safe, **without risk to health** and adequate as regards facilities and arrangements for their **welfare** at work.

Similar general and specific duties to protect health are placed on:

(a) employers with regard to people other than their employees (section 3);
(b) controllers of premises (section 4);
(c) manufacturers, designers, importers, etc. of articles and substances for use at work (section 6); and
(d) employees (section 7).

THE PRINCIPAL REGULATIONS

The principal regulations covering health-related issues are dealt with below. In many cases, these regulations are accompanied by an Approved Code of Practice (ACOP) and/or HSE Guidance Notes.

Control of Lead at Work Regulations 1980

These regulations, which are accompanied by an ACOP, are fundamentally concerned with the health protection of persons exposed to lead in its various forms while at work. **Lead** is defined in the regulations as meaning lead (including lead alloys and compounds and lead as a constituent of any substance) which is liable to be inhaled, ingested or absorbed by persons whether employees or not.

Where any work may expose anyone to lead, the employer must assess the risk before the work is commenced to determine the nature and degree of exposure. Adequate information, instruction and training must be provided for those persons likely to be affected.

Emphasis is placed on the introduction of measures to control materials, plant and processes so that adequate protection is provided against exposure to lead, thereby eliminating the need for the supply and use of RPE and PPE. Employers have a duty to ensure that any equipment for controlling exposure to lead is properly used and maintained, and employees must use these control measures properly. Employees have a duty further to report defects in the control measures to their employer. Where such control measures do not give adequate protection against airborne lead, approved respiratory protection must be provided and, where the exposure to lead is significant, each employee must be provided with adequate personal protective clothing.

Adequate washing facilities must be provided and arrangements made for the separate storage of personal clothing not worn during working hours and protective clothing. Workplaces must be kept clean and eating, drinking or smoking must be prohibited in places liable to be contaminated by lead.

Where the assessment of risk identifies the need for same, suitable air monitoring procedures must be implemented.

The principal duties of employers are covered in regulation 16 which covers medical surveillance procedures. Thus every employer must ensure that each of his employees who is employed on work which exposes the latter to lead is under medical surveillance by an employment medical adviser (EMA) or appointed doctor if:

(a) the exposure of that employee to lead is significant; or
(b) an EMA or appointed doctor certifies that the employee should be kept under medical surveillance.

In such cases, an employer must not expose that employee to lead, except in circumstances specified in the above certificate.

Every employee who is exposed to lead at work must, when required by the employer, present himself during normal working hours, for either of the following as may be necessary:

(a) medical examination;
(b) biological tests.

The regulations further require the keeping of records of any air monitoring, medical surveillance and testing, and maintenance of control measures.

The ACOP recommends that exposure to lead is significant if the level of airborne lead is in excess of half the lead-in-air standard or there is a substantial risk of ingesting lead or there is a risk of skin contact with concentrated lead alkyls. For lead-in-air the standard for an 8-hour time-weighted period is an average concentration of 0.15 mg/m^3 of air except for tetra-ethyl lead where the level is 0.10 mg/m^3 of air.

Health and Safety (First Aid) Regulations 1981

These regulations apply to nearly all workplaces in the UK. Under the regulations **first aid** means:

(a) in cases where a person will need help from a medical practitioner or nurse, treatment for the purpose of preserving life and minimising the consequences of injury or illness until such help is obtained; and
(b) treatment of minor injuries which would otherwise receive no treatment or which do not need treatment by a medical practitioner or nurse.

Duties of employers

Regulation 3 requires an employer to provide, or ensure that there are provided, such equipment and facilities as are **adequate and appropriate** in the circumstances for enabling first aid to be rendered to his employees if they are injured or become ill at work.

Two main duties are imposed on employers by the regulations:

(a) to provide first aid; and
(b) to inform employees of the first aid arrangements.

Self-employed persons must provide first aid equipment for their own use.

ACOP 1990

This ACOP emphasises the duty of employers to consider a number of factors and determine for themselves what is adequate and appropriate in all the circumstances. Furthermore, where there are particular risks associated with the operation of an enterprise, the employer must ensure that first aiders receive training to deal with these specific risks.

Factors to be considered in assessing first aid provision include:

(a) the number of employees;
(b) the nature of the undertaking;
(c) the size of the establishment and the distribution of employees;
(d) the location of the establishment and the locations to which employees go in the course of their employment;
(e) use of shift working (each shift would have the same level of first aid cover/protection); and
(f) the distance from external medical services e.g. local casualty department.

The general guidance suggests that even in a simple office there ought to be a first aider for every 50 persons.

First aid boxes

There should be at least one first aid box, the contents being listed in the ACOP.

Ionising Radiations Regulations 1985

These comprehensive regulations impose duties on employers to protect employees and other persons against ionising radiation arising from work with radioactive substances and other sources of ionising radiation. They also impose certain duties on employees.

The regulations provide an excellent framework for safe working practices involving the use of radioactive substances and sources of ionising radiation and the controls necessary. The principal requirements of the regulations are outlined below together with many of the definitions which are significant in the interpretation of the regulations.

Regulation 2 – Interpretation

Appointed doctor means a registered medical practitioner who is for the time being appointed in writing by the HSE for the purposes of these regulations.

Approved means approved for the time being by the HSE or the HSC, as the case may be.

Approved dosimetry service means a dosimetry service approved in accordance with regulation 9(1).

Classified person means a person who has been so designated in accordance with regulation 9(1).

Contamination means contamination by a radioactive substance of any surface (including any surface of the body or clothing) or any part of absorbent objects or materials or the contamination of liquids or gases by any radioactive substance.

Controlled area means an area which has been so designated by the employer in accordance with regulation 8(1) or (3).

Dose means, in relation to ionising radiation, any dose quantity or sum of dose quantities mentioned in Schedule 1.

Dose limit means, in relation to persons of a specified class, the dose limit specified in Schedule 1 in relation to a person of that class and is with respect to:

(a) the whole body, the relevant dose limit specified in Part I;
(b) any individual organ or tissue except the lens of the eye, the relevant dose limit specified in Part II;
(c) the lens of the eye, the relevant dose limit specified in Part III;
(d) the abdomen of a woman of reproductive capacity, the dose limit specified in Part IV; and
(e) the abdomen of a pregnant woman, the dose limit specified in Part V.

Dose rate means, in relation to a place, the rate at which a person or part of a person would receive a dose of ionising radiation from external radiation if he were at that place and **instantaneous dose rate** means a dose rate at that place averaged over one minute and **time average dose rate** means a dose rate at that place averaged over any 8 hour working period.

Dose record means, in relation to a person, the record of the doses received by that person as a result of his exposure to ionising radiation, being the record made and maintained on behalf of the employer by the approved dosimetry service in accordance with regulation 13(3)(a).

External radiation means, in relation to a person, ionising radiation coming from outside the body of that person.

Health record means, in relation to an employee, the record of medical surveillance of that employee maintained by the employer in accordance with regulation 16(c).

Internal radiation means, in relation to a person, ionising radiation coming from inside the body of that person.

Ionising radiation means gamma rays, X-rays or corpuscular radiations which are capable of producing ions either directly or indirectly.

Maintained, where the reference is to maintaining plant, apparatus or facilities, means maintained in an efficient state, in efficient working order and in good repair.

Medical exposure means exposure of a person to ionising radiation for the purpose of his medical or dental examination or treatment which is conducted under the direction of a suitably qualified person and includes any such examination or treatment conducted for the purposes of research.

Overexposure means any exposure of a person to ionising radiation to the extent that the dose received by that person causes a dose limit relevant to that person to be exceeded.

Qualified person means a person who has been so appointed for the purposes of regulation 24(3) in accordance with regulation 10(7).

Radiation generator means any apparatus in which charged particles are accelerated in a vacuum vessel through a potential difference of more than 5 kilovolts (whether or not in one or more steps) except an apparatus in which the only such generator is a cathode ray tube or visual display unit which does not cause, under normal operating conditions, an instantaneous dose rate of more than 5 μSvh^{-1} at a distance of 50mm from any accessible surface.

Radioactive substance means any substance having an activity concentration of more than 100 Bqg^{-1} and any other substance which contains one or more radionuclides whose activity cannot be disregarded for the purpose of radiation protection, and the term includes a radioactive substance in the form of a sealed source.

Radiation protection adviser means a person appointed in accordance with regulation 10(1).

Sealed source means a radioactive substance bonded wholly within a solid inactive material or encapsulated within an inactive receptacle of, in either case, sufficient strength to prevent any dispersion of the substance under reasonably foreseeable conditions of use and shall include the bonding or encapsulation, except that:

(a) where such bonding or encapsulation is solely for the purpose of storage, transport or disposal, the radioactive substance together with its bonding or encapsulation shall not be treated as a sealed source; and
(b) **sealed source** shall not include any radioactive substance inside a nuclear reactor or any nuclear fuel element.

Short-lived daughters of radon 222 means polonium 218, lead 214, bismuth 214 and polonium 214.

Supervised area means any area which has been so designated by the employer in accordance with regulation 8(2) or (3).

Trainee means a person aged 16 years or over (including a student) who is undergoing instruction or training which involves operations which would, in the case of an employee, be work with ionising radiation.

Woman of reproductive capacity means a woman who is made subject to the additional dose limit for a women of reproductive capacity specified in Part IV of Schedule 1 by an entry in her health record made by an employment medical adviser or appointed doctor.

Work with ionising radiation means any work:

(a) involving the production, processing, handling, use, holding, storage, moving, transport or disposal of any radioactive substance;
(b) involving the operation or use of any radioactive generator; or
(c) in which there is any exposure of a person to an atmosphere containing the short-lived daughters of radon 222 at a concentration in air, averaged over any 8-hour working period, of greater than 6.24×10^{-7} Jm^{-3} (0.03 working levels).

Working level means the special unit of potential alpha energy concentration in air and is any combination of short-lived daughters of radon 222 in unit volume of air such that the total potential alpha energy concentration for complete decay to lead 210 is 2.08×10^{-5} Jm^{-3}.

Principal provisions of the regulations

Part II – Dose limitation

Regulation 6 – Restriction of exposure

This regulation places a general duty on **employers** to restrict, so far as reasonably practicable, the exposure of employees and other persons foreseeably affected to ionising radiations.

In addition, every employer shall, so far as reasonably practicable, achieve this restriction of exposure to ionising radiation by means of engineering controls and design features which include shielding, ventilation, containment of radioactive substances and minimisation of contamination and, in addition, by the provision and use of safety features and warning devices.

Furthermore, in addition to taking the steps required above, every employer shall provide such systems of work as will, so far as reasonably practicable, restrict the exposure to ionising radiation of employees and other persons and, in the case of employees or other persons who enter or remain in controlled or supervised areas, provide those persons with adequate and suitable PPE (including RPE) unless:

(a) it is not reasonably practicable to further restrict exposure to ionising radiation by such means; or
(b) the use of PPE of a particular kind is not appropriate having regard to the nature of the work or the circumstances of the particular case.

An **employee** who is engaged in work with ionising radiation:

(a) shall not knowingly expose himself or any other person to ionising radiation to any extent greater than is reasonably necessary for the purposes of his work, and shall exercise reasonable care while carrying out such work;
(b) shall make full and proper use of any PPE provided; and
(c) shall forthwith report to his employer any defect he discovers in any PPE.

The **employer** shall ensure that:

(a) no radioactive substance in the form of a sealed source is held in the hand or manipulated directly by hand unless the instantaneous dose rate to the skin of the hand does not exceed 75 μSvh^{-1}; and
(b) so far as is reasonably practicable, no unsealed radioactive substance nor any article containing a radioactive substance is held in the hand or directly manipulated by hand.

No **employee** shall eat, drink, smoke, take snuff or apply cosmetics in any area which the employer has designated as a controlled area except that an employee may drink from a drinking fountain so constructed that there is no contamination of the water.

Regulation 7 – Dose limits

Under regulation 7 employers must ensure that employees and other persons are not exposed to ionising radiation to an extent that any dose limit specified in Schedule 1 for each such employee or other person, as the case may be, is exceeded.

Schedule 1 – Dose limits

Part I – Dose limits for the whole body

1 The dose limit for the whole body resulting from exposure to the whole or part of the body, being the sum of the following dose quantities resulting from exposure to ionising radiation, namely the effective dose equivalent from external radiation and the committed effect dose equivalent from that year's intake of radionuclides, shall in any calendar year be:

(a) for employees aged 18 years or over	50 mSv;
(b) for trainees aged under 18 years	15 mSv;
(c) for any other person	5 mSv.

Part II – Dose limits for individual organs and tissues

2 Without prejudice to Part I above, the dose limit for individual organs or tissues, being the sum of the following dose quantities resulting from exposure to ionising radiation, namely the dose equivalent from external radiation, the dose equivalent from contamination and the committed dose equivalent from that year's intake of radionuclides averaged throughout any individual organ or tissue (other than the lens of the eye) or any body extremity or over any area of skin, shall in any calendar year be:

(a) for employees aged 18 years or over 500 mSv;
(b) for trainees aged under 18 years 150 mSv;
(c) for any other person 50 mSv.

3 In assessing the dose quantity to skin whether from contamination or external radiation, the area of skin over which the dose quantity is averaged shall be appropriate to the circumstances but in any event shall not exceed 100 sq.cm.

Part III – Dose limits for the lens of the eye
4 The dose limit for the lens of the eye resulting from exposure to ionising radiation, being the average dose equivalent from external and internal radiation delivered between 2.5mm and 3.5mm behind the surface of the eye, shall in any calendar year be:

(a) for employees aged 18 years or over 150 mSv;
(b) for trainees aged under 18 years 45 mSv;
(c) for any other person 15 mSv.

Part IV – Dose limit for the abdomen of a woman of reproductive capacity
The dose limit for the abdomen of a woman of reproductive capacity who is at work, being the dose equivalent from external radiation resulting from exposure to ionising radiation averaged throughout the abdomen, shall be 13 mSv in any consecutive three-month interval.

Part V – Dose limit for the abdomen of a pregnant woman
The dose limit for the abdomen of a pregnant woman who is at work, being the dose equivalent from external radiation resulting from exposure to ionising radiation averaged throughout the abdomen, shall be 10 mSv during the declared term of pregnancy.

Note: The sievert (Sv) is the SI unit of dose equivalent, an index of the risk of harm from exposure of a particular body tissue to various radiations. The micro-sievert (mSv) is one thousandth of a sievert.

Part III – Regulation of work with ionising radiation

Part III lays down general procedures for ensuring safe working practices and the designation and appointment of certain persons.

Regulation 8 – Designation of controlled and supervised areas
Employers must designate as **controlled areas** any area where doses are likely to exceed three-tenths of any dose limit for employees of 18 years or over.

Employers must designate as a **supervised area** any area, not being a controlled area, where any person is likely to be exposed to an extent which exceeds one-third of the extent he would be exposed to in a controlled area.

The employer shall not permit any employee or other person to enter or remain in a controlled area unless that employee or other person:

(a) is a classified person; or
(b) enters or remains in the area under a written system of work (i.e. a permit to work system) such that:
   (i) in the case of an employee aged 18 years or over, he does not receive in any calendar year a dose of ionising radiation exceeding three-tenths of any relevant dose limit; or
   (ii) in the case of any other person, he does not receive in any calendar year a dose of ionising radiation exceeding any relevant dose limit.

An employer shall not permit an employee or other person to enter or remain in a controlled area in accordance with a written system of work unless he can demonstrate, by personal dose assessment or other measurements, that the doses are restricted in accordance with the above paragraph.

Regulation 9 – Designation of classified persons

The employer shall designate as **classified persons** those of his employees who are likely to receive a dose of ionising radiation which exceeds three-tenths of any relevant dose limit and shall forthwith inform those employees that they have been so designated.

An employer shall not designate an employee as a classified person unless:

(a) that employee is aged 18 years and over; and
(b) an employment medical adviser (EMA) or appointed doctor has certified in the health record that, in his professional opinion, that employee is fit to be designated as a classified person.

Regulation 10 – Appointment of radiation protection advisers and classified persons

In any case where:

(a) any employee is exposed to an instantaneous dose rate which exceeds 7.5 μSvh^{-1}; or
(b) the employer has designated a controlled area which people enter,

the employer carrying out work with ionising radiation shall appoint one or more **radiation protection advisers** (RPAs) for the purpose of advising him as to the observance of these regulations and as to other health and safety matters in connection with ionising radiation.

No employer shall appoint a person as an RPA unless:

(a) that person is suitably qualified and experienced;
(b) he has notified the HSE in writing of the intended appointment at least 28 days in advance; and
(c) he has received an acknowledgement in writing of the notification from the HSE.

The employer shall provide any RPA with adequate information and facilities for the performance of his functions and shall, whenever appropriate, consult that RPA.

The above provisions relating to qualifications, notification of the HSE and the provision of information and facilities, apply similarly in the case of a **qualified person** for the purpose of regulation 24 (monitoring of levels for radiation and contamination).

Regulation 11 – Local rules, supervision and radiation protection supervisors

Every employer who undertakes work with ionising radiation shall make and set down in writing local rules for the purpose of enabling the work with ionising radiation to be carried on in compliance with the requirements of these regulations and shall ensure that such of those rules as are relevant are brought to the attention of those employees and other persons who may be affected by them.

The employer shall ensure that the work with ionising radiation is supervised to the extent necessary to enable the work to be carried on in accordance with the requirements of these regulations and shall take all reasonable steps to ensure that any local rules that are relevant to that work are observed.

Where the work with ionising radiation is any such work other than that specified in Schedule 3 (Work not required to be notified under regulation 5(2)), the employer shall appoint one or more of his employees as **radiation protection supervisors** for the purpose of securing compliance with the above paragraph and any such appointment shall be in writing and the names of persons so appointed shall be included in the local rules.

Regulation 12 – Information, instruction and training

In addition to the general duty on employers under HSWA to inform, instruct and train employees and other persons, this regulation specifies groups of people who must receive information, instruction and training.

Every employer shall ensure that:

(a) those of his **employees** who are engaged in work with ionising radiation receive such information, instruction and training as will enable them to conduct the work in accordance with the requirements of these regulations;

(b) adequate information is given to **other persons** who are directly concerned with the work with ionising radiation carried on by the employer to ensure their health and safety, so far as is reasonably practicable;

(c) **classified persons and trainees** are informed of the health hazard, if any, associated with their work, the precautions to be taken and the importance of complying with the medical and technical requirements and are given appropriate training in the field of radiation protection; and

(d) those of his employees who are engaged in work with ionising radiation and who are **women** are informed of the possible hazard arising from ionising radiation to the foetus in early pregnancy and of the importance of informing the employer as soon as they discover that they have become pregnant.

Part IV – Dosimetry and medical surveillance

This Part deals with:

(a) the procedures for dose assessment (regulation 13);
(b) dosimetry in the event of an accident, occurrence or incident which is likely to result in a person being exposed to ionising radiation to an extent greater than three-tenths of any relevant dose limit (regulation 14);
(c) medical surveillance of certain groups (regulation 16); and
(d) approved arrangements for the protection of certain employees (regulation 17).

Part V – Arrangements for the control of radioactive substances

Regulation 18 – Sealed sources and articles containing or embodying radioactive substances

Regulation 18 imposes duties on employers with regard to the control of radioactive substances.

Where a radioactive substance is used as a source of ionising radiation in work with ionising radiation, the employer shall ensure that, whenever reasonably practicable, the substance is in the form of a sealed source.

The employer shall ensure that the design, construction and maintenance of any article containing or embodying a radioactive substance, including its bonding, immediate container or other mechanical protection, is such as to prevent the leakage of any radioactive substance:

(a) in the case of a sealed source, so far as is practicable; or
(b) in the case of any other article, so far as is reasonably practicable.

Where appropriate, the employer shall ensure that suitable tests are carried out at suitable intervals, which shall in no case exceed 26 months, to detect leakage of radioactive substances from any article, and the employer shall keep a suitable record of those tests for at least three years from the date of the tests to which they refer.

Regulation 19 – Accounting for radioactive substances

This regulation requires an employer to take such steps as are appropriate to account for and keep records of the quantity and location of radioactive substances, and to keep records for at least two years from that date and at least two years from the date of disposal of a radioactive substance.

Regulation 20 – Keeping of radioactive substances
Every employer shall ensure, so far as is reasonably practicable, that any radioactive substance under his control which is not for the time being in use or being moved, transported or disposed of:

(a) is kept in a suitable receptacle; and
(b) is kept in a suitable store.

Regulation 21 – Transport and moving of radioactive substances
Every employer who causes or permits a radioactive substance to be transported shall ensure that, so far as is reasonably practicable, the substance is kept in a suitable receptacle, suitably labelled, while it is being transported.

So far as is reasonably practicable, any written information accompanying the radioactive substance shall enable a person receiving it:

(a) to know the nature and quantity of the radioactive substance; and
(b) to comply with the requirements of regulations 6, 7, 8 and 18(2) and the above paragraph of this regulation in so far as such compliance depends on that information.

Regulation 22 – Washing and changing facilities
Where an employer is required to designate an area as a controlled area or a supervised area, he shall ensure that adequate washing and changing facilities are provided for persons who enter or leave that area and that those facilities are properly maintained.

Regulation 23 – Personal protective equipment (PPE)
Every employer shall ensure that, where PPE provided includes respiratory protective equipment (RPE), that RPE is of a type, or conforms to a standard, approved in either case by the HSE.

Every employer shall ensure that all PPE (including RPE) is thoroughly examined at suitable intervals and is properly maintained and that, in the case of RPE, a suitable record of that examination is made and kept for at least two years from the date on which the examination was made and that the record includes a statement of the condition of the equipment at the time of the examination.

Part VI – Monitoring of ionising radiation

Regulation 24 – Monitoring of levels for radiation and contamination
Regulation 24 places the following duties on the employer:

(a) to take such steps, otherwise than by the use of assessed doses of individuals, to ensure levels of radiation are adequately monitored for each controlled area or supervised area;
(b) to provide equipment for such monitoring which:

(i) is properly maintained;
(ii) is thoroughly examined and tested at least once in every 14 months; and
(iii) has had its performance established by tests before being taken into use for the first time;

(c) to ensure that the above examinations and tests are carried out by, or under the immediate supervision of, a **qualified person**; and
(d) to make suitable records of the results of the monitoring and the tests carried out and to keep such records for at least two years from the respective dates on which they were made.

Part VII – Assessments and notifications

Regulation 25 – Assessment of hazards

Before work with ionising radiation is carried on, an employer must make an assessment which is adequate to identify the nature and magnitude of the radiation hazard to employees which is likely to arise from that work in the event of any reasonably foreseeable accident, occurrence or incident.

Where such an assessment shows that a radiation hazard to employees or other persons exists, the employer shall take all reasonably practicable steps to:

(a) prevent any such accident, occurrence or incident;
(b) limit the consequences of any such accident, occurrence or incident which does occur; and
(c) provide employees with the information, instruction and training and with the equipment necessary to restrict their exposure to ionising radiation.

Regulation 26 – Special hazard assessments and reports

An employer shall not undertake any work with ionising radiation which involves:

(a) having on any site;
(b) providing facilities for there to be on any site; or
(c) transporting,

more than the quantity of any radioactive substance specified in column 6 of Schedule 2 or, in the case of fissile material, the mass of that material specified below, unless he has made an assessment of the radiation hazard that could arise from that work and has sent a report of the assessment to the HSE at least 28 days before commencing that work or within such shorter time in advance as the HSE may agree.

The specified mass of fissile material shall be:

(a) plutonium as Pu 239 or as a mixture of plutonium isotopes containing Pu 239 or Pu 241, 150 grams;
(b) uranium as U 233, 150 grams;

(c) uranium enriched in U 235 to more than 1 per cent but not more than 5 per cent, 500 grams; and
(d) uranium enriched in U 235 to more than 5 per cent, 250 grams..

Regulation 27 – Contingency plans

Where the assessment made in accordance with regulation 25 shows that as a result of any reasonably foreseeable accident, occurrence or incident:

(a) employees or other persons are likely to receive a dose of ionising radiation which exceeds any relevant dose limit; or
(b) any area other than a controlled area would be required to be so designated

the employer shall prepare a contingency plan designed to secure, so far as is reasonably practicable, the restriction of exposure to ionising radiation and the health and safety of persons who may be affected by the accident, occurrence or incident to which the plan relates.

Regulation 28 – Investigation of exposure

The employer shall ensure that an investigation is carried out forthwith when any of his employees is exposed to ionising radiation to an extent that three-tenths of the annual dose limit for employees aged 18 years or over specified in Part 1 to Schedule 1 (whole body dose limit) is exceeded for the first time in any calendar year, to determine whether the requirements of regulation 6 are being met. A report of any such investigation shall be kept for at least two years.

Regulation 29 – Investigation and notification of exposure

This regulation lays down procedures for notifying the HSE and making a detailed investigation where an employer suspects or has been informed that an employee is likely to have received an overexposure.

Regulation 30 – Notification of certain occurrences

Every employer shall forthwith notify the HSE in any case where a quantity of radioactive substance which was under his control and which exceeds the quantity specified for that substance in column 7 of Schedule 2:

(a) has been released or is likely to have been released into the atmosphere as a gas, aerosol or dust; or
(b) has been spilled or otherwise released in such a manner as to give rise to significant contamination.

The employer shall make an immediate investigation and, unless no occurrence has occurred, notify the HSE. A report of this investigation shall be kept for at least 50 years from the date on which it was made.

Part VIII – Safety of articles and equipment

Regulation 32 – Duties of manufacturers etc. of articles for use in work with ionising radiation

This regulation modifies the duties of manufacturers etc. under section 6 of HSWA to include a duty to ensure that any such article is so designed and constructed as to restrict, so far as is reasonably practicable, the extent to which employees and other persons are or are likely to be exposed to ionising radiation.

Specific duties are also placed on persons who erect or install an article for use at work. Such persons shall:

(a) where appropriate, together with an RPA, undertake a critical examination of the way in which the article was erected or installed for the purpose of ensuring, in particular, that:
  (i) the safety features and warning devices operate correctly; and
  (ii) there is sufficient protection for persons from exposure to ionising radiation; and
(b) provide the employer with adequate information about proper use, testing and maintenance of the article.

Regulation 33 – Equipment used for medical exposure

Every employer shall ensure that any equipment or apparatus under his control which is used in connection with medical exposure is of such design or construction and is so installed and maintained as to be capable of restricting, so far as is reasonably practicable, the exposure to ionising radiation of any person who is undergoing a medical exposure to the extent that this is compatible with the clinical purpose or research objective in view.

The employer must investigate any incident where a person has been overexposed as a result of a malfunction of, or a defect in, such radiation equipment and, unless that investigation shows beyond reasonable doubt that no such incident has occurred, shall forthwith notify the HSE. He shall then make a detailed investigation of the circumstances of exposure and an assessment of the dose received. Such a report shall be kept for at least 50 years from the date on which it was made.

Regulation 34 – Misuse of or interference with sources of ionising radiation

No person shall intentionally or recklessly misuse or without reasonable excuse interfere with any radioactive substance or radiation generator.

Part IX – Miscellaneous and general

Regulation 35 – Defence on contravention of certain regulations

It shall be a defence in proceedings against any person for an offence consisting of a contravention of regulations 5(4), 17(1) and 26(4) for the person to prove that at the time proceedings were commenced:

(a) an improvement notice under section 21 of HSWA relating to the contravention had not been served on him; or
(b) if such a notice had been served on him:
    (i) the period for compliance had not expired; or
    (ii) he had appealed against the notice and that appeal had not been dismissed or withdrawn.

Schedules to the regulations

The ten Schedules to the regulations cover the following matters:

1 Dose limits
2 Quantities of radionuclides
   Part I Table of radionuclides
   Part II Quantity ratios for more than one radionuclide
3 Work not required to be notified under regulation 5(2)
4 Particulars to be supplied in a notification under regulation 5(2)
5 Additional particulars that the HSE may require
6 Designation of controlled areas
   Part I Designation in relation to external radiation
   Part II Designation in relation to internal radiation
   Part III Designation in relation to external radiation and internal radiation together
   Part IV Designation in relation to short-lived daughters of radon 222
7 Particulars to be included in an assessment report
8 Further particulars that the HSE may require
9 Sealed sources to which regulation 26 does not apply
10 Revocations and modifications.

Reporting and recording of injuries, diseases and dangerous occurrences

Under the Reporting of Injuries, Diseases and Dangerous Occurrences Regulations 1985, a 'responsible person' must notify, by quickest practicable means e.g. telephone or fax, and report within seven days to the enforcing authority certain classes of injury sustained by people at work, various occupational diseases and defined 'dangerous occurrences' as listed in the Schedule to the regulations.

In these regulations, the following definitions are significant:

Specified major injury or condition

This means:

(a) fracture of the skull, spine or pelvis;
(b) fracture of any bone:
    (i) in the arm or wrist, but not a bone in the hand;

    (ii) in the leg or ankle, but not a bone in the foot;
(c) amputation of:
    (i) a hand or foot;
    (ii) a finger, thumb or toe, or any part thereof, if the joint or bone is completely severed;
(d) the loss of sight of an eye, a penetrating injury to the eye, or a chemical or hot metal burn to the eye;
(e) either injury (including burns) requiring immediate medical treatment, or loss of consciousness, resulting in either case from an electric shock from any electrical circuit or equipment, whether or not due to direct contact;
(f) loss of consciousness resulting from lack of oxygen;
(g) decompression sickness (unless suffered during an operation to which the Diving Operations at Work Regulations 1981 apply) requiring immediate medical treatment;
(h) either acute illness requiring medical treatment, or loss of consciousness, resulting in either case from absorption of any substance by inhalation, ingestion or through the skin;
(j) acute illness requiring medical treatment where there is reason to believe this resulted from exposure to a pathogen or infected material; and
(k) any other injury which results in the person injured being admitted immediately into hospital for more than 24 hours.

Reportable disease

This is a disease specified in column 1 of Schedule 2 of the regulations and involving one of the activities specified in the corresponding entry in column 2 of that Schedule.

Examples:

Column 1 *Disease*	*Column 2* *Work activity*
2 Chrome ulceration of: (a) the nose or throat; or (b) the skin of the hands or forearm	Work involving exposure to chromic acid or to any other chromium compound
10 Byssinosis	Work in any room where any process up to and including the weaving process is performed in a factory in which the spinning or manipulation of raw or waste cotton or of flax, or the weaving of cotton or flax, is carried on
16 Hepatitis	Work involving exposure to human blood products or body secretions or excretions

The responsible person must forthwith send a report to the enforcing authority on the approved form (Form 2508A) wherever a person suffers from one of these diseases.

Dangerous occurrence
This is an occurrence which arises out of or in connection with work and is of a class specified in Part 1 of Schedule 1 of the regulations.

Examples:

3 Explosion, collapse or bursting of any pressure vessel, including a boiler or boiler tube, in which the internal pressure was above or below atmospheric pressure, which might have been liable to cause the death of, or any of the injuries or conditions covered by regulation 3(2) to, any person, or which resulted in stoppage of the plant involved for more than 24 hours.

7 A collapse or partial collapse of any scaffold which is more than 5 metres high which results in a substantial part of the scaffold falling or overturning; and where the scaffold is slung or suspended, a collapse or partial collapse of the suspension arrangements (including any outrigger) which causes a working platform or cradle to fall more than 5 metres.

10 Any ignition or explosion of explosives, where the ignition or explosion was not intentional.

Responsible person
In the case of a reportable injury or disease:

(a) involving an employee at work – the employer; and
(b) involving a person undergoing training for employment – the person whose undertaking makes the immediate provision of that training.

In any other case, the person for the time being having control of the premises in connection with the carrying on by him of any trade, business or other undertaking (whether for profit or not) at which, or in connection with the work at which, the accident, disease or dangerous occurrence happened.

Work
This means work as an employee, as a self-employed person or as a person undergoing training for employment (whether or not under any scheme administered by the Manpower Services Commission).

Notification and reporting of injuries and dangerous occurrences
Where any person, as a result of an accident arising out of or in connection with work, dies or suffers any of the specific major injuries or conditions, or where there is a dangerous occurrence, the responsible person shall:

(a) forthwith notify the enforcing authority by quickest practicable means; and
(b) within seven days send a report thereof to the enforcing authority on a form approved for this purpose (Form 2508).

Where a person at work is incapacitated for work of a kind which he might reasonably be expected to do, either under his contract of employment, or, if there is no such contract, in the normal course of his work, for more than three consecutive days (excluding the day of the accident but including any days which would not have been working days) because of an injury (other than a specified major injury) resulting from an accident at work, the responsible person shall within seven days of the accident send a report thereof to the enforcing authority on the form approved for this purpose (Form 2508).

Reporting the death of an employee

Where an employee, as a result of an accident at work, has suffered a specified major injury or condition which is the cause of his death within one year of the date of the accident, the employer shall inform the enforcing authority in writing of the death as soon as it comes to his knowledge, whether or not the accident had been reported previously.

Reporting a case of disease

Where a person at work suffers from a reportable disease, the responsible person shall forthwith send a report thereof to the enforcing authority on the form approved for this purpose (Form 2508A). This requirement applies only if:

(a) in the case of an employee or a person undergoing training, the responsible person has received a written statement prepared by a registered medical practitioner diagnosing the disease as one of the reportable diseases; or
(b) in the case of a self-employed person, that person has been informed by a registered medical practitioner that he is suffering from the disease so specified.

Records

The responsible person shall keep a record of:

(a) any event which is required to be reported under regulation 3 i.e. specified major injury or condition or dangerous occurrence; and
(b) any case of disease required to be reported under regulation 3.

The records shall be kept at the place where the work to which they relate is carried on or, if that is not reasonably practicable, at the usual place of business of the responsible person and an entry in either of such records shall be kept for at least three years from the date on which it is made.

The responsible person shall send to the enforcing authority such extracts from the records required to be kept as the enforcing authority may from time to time require.

In the case of (a) above, the following particulars shall be kept in records of any event which is reportable under regulation 3:

1 Date and time of the accident or dangerous occurrence.
2 The following particulars of the person affected:
   (a) full name;
   (b) occupation;
   (c) nature of the injury or condition.
3 Place where the accident or dangerous occurrence occurred.
4 A brief description of the circumstances.

In the case of (b) above, the following records shall be kept of instances of the diseases specified in Schedule 2 and reportable under regulation 5:

1 Date of diagnosis of the disease.
2 Occupation of the person affected.
3 Name and nature of the disease.

Defence in proceedings for an offence contravening the regulations

It shall be defence in proceedings against any person for an offence under these regulations for that person to prove that he was not aware of the event requiring him to notify or send a report to the enforcing authority and that he had taken all reasonable steps to have all such events brought to his notice.

Accident books

Under the Social Security Act 1975:

1 Employers must notify their employer of any accidents resulting in personal injury in respect of which benefit may be payable. Notification may be given by a third party if the employee is incapacitated.
2 Employees must enter the appropriate particulars of all accidents in an Accident Book (Form BI 510). This may be done by another person if the employee is incapacitated. Such an entry is deemed to satisfy the requirements of 1 above.
3 Employers must investigate all accidents of which notice is given by employees. Variations in the findings of this investigation and the particulars given in the notification must be recorded.
4 Employers must, on request, furnish the Department of Social Security with such information as may be required relating to accidents in respect of which benefit may be payable e.g. Forms 2508 and 2508A.
5 Employers must provide and keep readily available an Accident Book in an approved form in which the appropriate details of all accidents can be recorded (Form BI 510). Such books, when completed, should be retained for three years after the date of the last entry.
6 For the purpose of the above, the appropriate particulars should include:
   (a) Name and address of injured person.
   (b) Date and time of accident.
   (c) Place where accident happened.
   (d) Cause and nature of injury.
   (e) Name and address of any third party giving notice.

Social Security (Industrial Injuries) (Prescribed Diseases) Regulations 1985

These regulations list those diseases which are prescribed for the purpose of payment of disablement benefit. They were subject to minor amendments in 1987 and 1989.

A prescribed disease is defined in the Social Security Act 1975 as:

(a) a disease which ought to be treated, having regard to its causes and incidence and other relevant considerations, as a risk of occupation and not a risk common to all persons; and
(b) it is such that, in the absence of special circumstances, the attribution of particular cases to the nature of the employment can be established with reasonable certainty.

Schedule 1 to the regulations classifies prescribed diseases or injuries thus:

(a) conditions due to physical agents;
(b) conditions due to biological agents;
(c) conditions due to chemical agents; and
(d) miscellaneous conditions.

Within these four classifications, prescribed diseases or injuries are related to specific occupations. Each disease is numbered within the particular classification.

The Schedule is very comprehensive and examples are listed in the four classifications given below.

Prescribed disease or injury	*Occupation* *Any occupation involving:*
A Conditions due to physical agents	
2 Heat cataract	Frequent or prolonged exposure to rays from molten or red-hot material
5 Subcutaneous cellulitis of the hand (Beat hand)	Manual labour causing severe or prolonged friction or pressure on the hand
9 Miner's nystagmus	Work in or about a mine
B Conditions due to biological agents	
1 Anthrax	Contact with animals infected with anthrax or the handling (including the loading or unloading or transport) of animal products or residues
5 Tuberculosis	Work in or about a mine

9 Infection by Streptococcus suis	Contact with pigs infected with Streptococcus suis, or with the carcasses, products or residues of pigs so infected
C Conditions due to chemical agents	
5 Poisoning by mercury or compound of mercury	The use or handling of, or exposure to the fumes, dust or vapour of, mercury or a compound of mercury, or a substance containing mercury
21 (a) Localised new growth of the skin, papillomatous or keratotic (b) Squamous-celled carcinoma of the skin	The use or handling of, or exposure to, arsenic, tar, pitch, bitumen, mineral oil (including paraffin), soot or any compound, product or or residue of any of these substances, except quinone or hydroquinone
26 Damage to the liver or kidneys due to exposure to carbon tetrachloride	The use of or handling of or exposure to the fumes of, or vapour containing carbon tetrachloride
D Miscellaneous conditions	
2 Byssinosis	Work in any room where any process up to and including the weaving process is performed in a factory in which the spinning or manipulation of raw or waste cotton, or of flax, or the weaving of cotton or flax, is carried on
8 Primary carcinoma of the lung where there is accompanying evidence of one or both of the following (a) asbestosis; (b) bilateral diffuse pleural thickening	(a) The working or handling of asbestos or any admixture of asbestos; or (b) the manufacture or repair of asbestos textiles or other articles containing or composed of asbestos; or (c) the cleaning of any machinery or plant used in any of the foregoing operations and of any chambers, fixtures or appliances for the collection of asbestos dust; or

9

	(d) substantial exposure to the dust arising from any of the foregoing operations
10 Lung cancer	(a) Work underground in a tin mine; or (b) exposure to bis (chloromethyl) ether produced during the manufacture of chloromethyl methyl ether; or (c) exposure to pure zinc chromate, calcium chromate or strontium chromate

Under the regulations, pneumoconiosis, and the various causes of same, are treated separately.

Control of Asbestos at Work Regulations 1987

These regulations apply to all workplaces where asbestos products are manufactured, used or handled, and to all persons at risk from work with asbestos, and give specific statutory protection to all those who may be affected by work activities involving asbestos. They are accompanied by an ACOP, 'The Control of Asbestos at Work', and a series of HSE Guidance Notes.

The principal requirements of the regulations are as follows:

(a) a general duty on employers to prevent exposure of employees; where this is not reasonably practicable, employers must reduce the exposure by means other than the use of respiratory protective equipment (RPE) (regulation 3);
(b) prohibitions on work involving asbestos (regulation 4);
(c) health risk assessment prior to undertaking work that exposes or is liable to expose employees (regulation 5);
(d) notification to the enforcing authority at least 28 days beforehand prior to commencing work involving asbestos (regulation 6);
(e) the provision of information, instruction and training for employees liable to be affected by exposure (regulation 7);
(f) prevention/control of employee exposure to asbestos above specified action levels (regulation 8);
(g) duties on employers to ensure PPE is properly used or applied, on employees to make full and proper use of any control measures, PPE or other facilities, and to report defects in control measures and PPE (regulation 9);
(h) a duty on employers to ensure RPE is maintained, examined and tested by a competent person (regulation 10);

(i) measures to ensure the provision, disposal or cleaning of protective clothing (regulation 11);
(j) measures to be taken by employers to prevent or reduce the spread of asbestos to the lowest level reasonably practicable (regulation 12);
(k) a duty on employers to maintain work premises in a clean condition (regulation 13);
(l) the designation of asbestos exposure areas and respirator zones (regulation 14);
(m) a duty to undertake air monitoring (regulation 15);
(n) a duty to maintain health records and records of employee exposures (regulation 15);
(o) measures to ensure the provision of appropriate welfare facilities, i.e. washing facilities, clothing and RPE storage facilities (regulation 17);
(p) containment of asbestos waste in suitable and sealed containers, suitably labelled and marked, during storage, reception into premises or distribution (regulation 18); and
(q) compulsory labelling/marking of asbestos products (regulation 19).

Control of Substances Hazardous to Health (COSHH 2) Regulations 1994

The COSHH Regulations apply to all forms of workplace and every form of work activity involving the use of substances which may be hazardous to health to people at work. The Regulations are supported by a number of Approved Codes of Practice (ACOPS), including 'Control of Substances Hazardous to Health', 'Control of Carcinogenic Substances' and 'Control of Biological Agents'. Because of the relative significance of these regulations to all who use or come into contact with substances hazardous to health at work, the full extent of the duties under the regulations are covered below.

Introduction to the COSHH Regulations

The regulations and the various ACOPs set out a strategy for safety with substances hazardous to health covering more than 40,000 chemicals and materials, together with hazardous substances generated by industrial processes.

The strategy established in the COSHH Regulations covers four main areas.

(a) acquisition and dissemination of information and knowledge about hazardous substances;
(b) the assessment of risks to health associated with the use, handling, storage, etc. of such substances at work;
(c) elimination or control of health risks by the use of appropriate engineering applications, operating procedures and personal protection;
(d) monitoring the effectiveness of the measures taken.

It should be appreciated that the majority of the duties imposed on employers and others are of an absolute or strict nature. The 1994 regulations revoke in full the 1988 regulations.

The COSHH Regulations 1994

Application of regulations 6 to 12

1 Regulations 6 to 12 shall have effect with a view to protecting persons against risks to their health, whether immediate or delayed, arising from exposure to substances hazardous to health **except**:
   (a) lead – so far as the Lead at Work Regulations 1980 apply; and asbestos – so far as the Control of Asbestos at Work Regulations 1987 apply;
   (b) where the substance is hazardous solely by virtue of its radioactive, explosive or flammable properties, or solely because it is at a high or low temperature or a high pressure;
   (c) where the risk to health is a risk to the health of a person to whom the substance is administered in the course of his medical treatment;
   (e) below ground in any mine within the meaning of section 180 of the Mines and Quarries Act 1954.

2 In paragraph (1)(c) **medical treatment** means medical or dental examination or treatment which is conducted by, or under the direction of, a registered medical practitioner or registered dentist and includes any such examination, treatment or administration of any substance conducted for the purpose of research.

3 Nothing in these Regulations shall prejudice any requirement imposed or under any enactment relating to public health or the protection of the environment.

Regulation 2 – Definitions

The following definitions in Regulation 2 are of significance:

Approved supply list has the meaning assigned to it in regulation 4(1) of the Chemicals (Hazard Information and Packaging) Regulations 1993

Biological agent means any micro-organism, cell, culture, or human endoparasite, including any which have been genetically modified, which may cause any infection, allergy, toxicity or otherwise create a hazard to human health.

Carcinogen means

(a) any substance or preparation which if classified in accordance with the classification provided by regulation 5 of the Chemicals (Hazard Information and Packaging) Regulations 1993 would be in the category 6 danger, carcinogenic (category 1) or carcinogenic (category 2) whether or not the substance or preparation would be required to be classified under those Regulations; or

(b) any substance or preparation
  (i) listed in Schedule 8; and
  (ii) any substance or preparation arising from a process specified in Schedule 8 which is a substance hazardous to health.

Fumigation means any operation in which a substance is released into the atmosphere so as to form a gas to control or kill pests or other undesirable organisms; and **fumigate** and **fumigant** shall be construed accordingly.

Maximum exposure limit for a substance hazardous to health means the maximum exposure limit for that substance set out in Schedule 1 in relation to the reference period specified therein when calculated by a method approved by the HSC.

Micro-organism includes any microbiological entity, cellular or non-cellular, which is capable of replication or of transferring genetic material.

Occupational exposure standard for a substance hazardous to health means the standard approved by the HSC for that substance in relation to the specified reference period when calculated by a method approved by the HSC.

Substance means any natural or artificial substance whether in solid or liquid form or in the form of a gas or vapour (including micro-organisms).

Substance hazardous to health means any substance (including any preparation) which is:

(a) a substance which is listed in Part 1 of the approved supply list as dangerous for supply within the meaning of the Chemicals (Hazard Information and Packaging) Regulations 1993 and for which an indication of danger specified for the substance in Part V of that list is very toxic, toxic, harmful, corrosive or irritant;
(b) a substance specified in Schedule 1 (which lists substances assigned maximum exposure limits) or for which the HSC has approved an occupational exposure standard (see current HSE Guidance Note EH40 'Occupational Exposure Limits');
(c) a biological agent;
(d) dust of any kind when present at a substantial concentration in air; and
(e) a substance, not being a substance mentioned in sub-paragraphs (a) to (d) above, which creates a hazard to the health of any person which is comparable with the hazards created by substances mentioned in those sub-paragraphs.

Regulation 6 – Assessment of health risks created by work involving substances hazardous to health

1 An employer shall not carry on any work which is liable to expose any employees to any substance hazardous to health unless he has made a **suitable and sufficient assessment** of the risks created by that work to the health of those employees and of the steps that need to be taken to

meet the requirements of the regulations.

2 This assessment shall be reviewed forthwith if
   (a) there is reason to suspect that the assessment is no longer valid; or
   (b) there has been a significant change in the work to which the assessment relates;

   and, where as a result of the review, changes in the assessment are required, those changes shall be made.

Note

A 'suitable and sufficient' assessment

The General ACOP indicates that a suitable and sufficient assessment should include:

(a) an assessment of the risks to health;
(b) the steps which need to be taken to achieve adequate control of exposure, in accordance with regulation 7; and
(c) identification of other action necessary to comply with regulations 8–12.

An assessment of the risks created by any work should involve:

(a) a consideration of:
   (i) which substances or types of substances (including micro-organisms) employees are liable to be exposed to (taking into account the consequences of possible failure of any control measures provided to meet the requirements of regulation 7);
   (ii) what effects those substances can have on the body;
   (iii) where the substances are likely to be present and in what form;
   (iv) the ways in which and the extent to which any groups of employees or other persons could potentially be exposed, taking into account the nature of the work and process, and any reasonably foreseeable deterioration in, or failure of, any control measure provided for the purpose of regulation 7;
(b) an estimate of exposure, taking into account engineering measures and systems of work currently employed for controlling potential exposure;
(c) where valid standards exist, representing adequate control, comparison of the estimate with those standards.

Detailed guidance on COSHH assessment is provided in the HSE publication 'A Step by Step Guide to COSHH Assessment' (HMSO).

Regulation 7 – Prevention or control of exposure to substances hazardous to health

1 Every employer shall ensure that the exposure of his employees to substances hazardous to health is either prevented or, where this is not reasonably practicable, adequately controlled.

2 So far as is reasonably practicable, the prevention or adequate control of exposure of employees to substances hazardous to health, except to a carcinogen or biological agent, shall be secured by measures other than the provision of personal protective equipment.

3 Without prejudice to the generality of paragraph 1, where the assessment made under regulation 6 shows that it is not reasonably practicable to prevent exposure to a **carcinogen** by using an alternative substance or process, the employer shall employ **all** the following measures, namely:
  (a) the total **enclosure** of the process and handling systems unless this is not reasonably practicable;
  (b) the use of plant, processes and systems of work which **minimise** the generation of, or **suppress and contain**, spills, leaks, dust, fumes and vapours of carcinogens;
  (c) the **limitation** of the quantities of a carcinogen at the place of work;
  (d) the keeping of the number of persons exposed to a **minimum;**
  (e) the prohibition of **eating, drinking and smoking** in areas that may be contaminated by carcinogens;
  (f) the provision of **hygiene measures** including adequate washing facilities and regular cleaning of walls and surfaces;
  (g) the **designation** of those areas and installations which may be contaminated by carcinogens, and the use of suitable and sufficient **warning signs**; and
  (h) the **safe storage, handling and disposal** of carcinogens and the use of closed and clearly labelled **containers.**

4 Where the measures taken in accordance with paragraphs 2 or 3, as the case may be, do not prevent, or provide adequate control of, exposure to substances hazardous to health to which those paragraphs apply, then, **in addition** to taking those measures, the employer shall provide those employees with suitable **personal protective equipment** as will adequately control their exposure to those substances.

5 Any personal protective equipment provided by an employer shall comply with any enactment which implements in Great Britain any provision on design or manufacture with respect to health or safety in any relevant Community directive listed in Schedule 1 to the Personal Protective Equipment at Work Regulations 1992 which is applicable to that item of personal protective equipment.

6 Where there is exposure to a substance for which a maximum exposure limit (MEL) is specified in Schedule 1, the control of exposure shall, so far as the inhalation of that substance is concerned, only be treated as being adequate if the level of exposure is reduced so far as is reasonably practicable and in any case below the MEL.

7 Without prejudice to the generality of paragraph 1, where there is exposure to a substance for which an occupational exposure standard (OES) has been approved, the control of exposure shall, so far as the inhalation of the substance is concerned, only be treated as adequate if:
  (a) the OES is not exceeded; or
  (b) where the OES is exceeded, the employer identifies the reasons for the standard being exceeded and takes appropriate action to remedy the situation as soon as is reasonably practicable.

8 Where **respiratory protective equipment** is provided in pursuance of this regulation, then it shall:
   (a) be suitable for the purpose; and
   (b) comply with paragraph 5 or, where no requirement is imposed by virtue of that paragraph, be of a type approved or shall conform to a standard approved, in either case, by the HSE.
9 In the event of a failure of a control measure which might result in the escape of **carcinogens** into the workplace, the employer shall ensure that:
   (a) only those persons who are responsible for the carrying out of repairs and other necessary work are permitted in the affected area and they are provided with suitable respiratory protective equipment and protective clothing; and
   (b) employees and other persons who may be affected are informed of the failure forthwith.
10 Schedule 9 of these Regulations shall have effect in relation to biological agents.
11 In this regulation **adequate** means adequate having regard only to the nature of the substance and the nature and degree of exposure to substances hazardous to health and **adequately** shall be construed accordingly.

Regulation 8 – Use of control measures, etc.

1 Every **employer** who provides any control measure, PPE or other thing or facility pursuant to these regulations shall take all reasonable steps to ensure that it is properly used or applied as the case may be.
2 Every **employee** shall make full and proper use of any control measure, PPE or other thing or facility provided pursuant to these regulations and shall take all reasonable steps to ensure it is returned after use to any accommodation provided for it and, if he discovers any defect therein, he shall report it forthwith to his employer.

Regulation 9 – Maintenance, examination and test of control measures, etc.

1 Any employer who provides any control measure to meet the requirements of regulation 7 shall ensure that it is maintained **in efficient state, in efficient working order and in good repair** and, in the case of personal protective equipment, in a clean condition.
2 Where engineering controls are provided to meet the requirements of regulation 7, the employer shall ensure that thorough examinations and tests of those engineering controls are carried out:
   (a) in the case of local exhaust ventilation (LEV) plant, at least once every 14 months, or for LEV plant used in conjunction with a process specified in column 1 of Schedule 3, at not more than the interval specified in the corresponding entry in column 2 of that Schedule; and
   (b) in any other case, at suitable intervals.
3 Where respiratory protective equipment (RPE) (other than disposable

RPE) is provided to meet the requirements of regulation 7, the employer shall ensure that at suitable intervals thorough examinations and, where appropriate, tests of that equipment are carried out.

4 Every employer shall keep a suitable record of examinations and tests carried out in accordance with the paragraphs 2 and 3, and of any repairs carried out as a result of those examinations and tests, and that record or a suitable summary thereof, shall be kept available for at least five years from the date on which it was made.

Regulation 10 – Monitoring exposure at the workplace

1 In any case in which:
   (a) it is a requisite for ensuring the maintenance of adequate control of the exposure of employees to substances hazardous to health; or
   (b) it is otherwise requisite for protecting the health of employees, the employer shall ensure that the exposure of employees to substances hazardous to health is monitored in accordance with a suitable procedure.

2 Where a substance or process is specified in column 1 of Schedule 4, monitoring shall be carried out at the frequency specified in the corresponding entry in column 2 of that Schedule.

3 The employer shall keep a record of any monitoring carried out for the purpose of the regulation and that that record or a suitable summary thereof shall be kept available:
   (a) where the record is representative of the personal exposure of identifiable employees, for at least 40 years;
   (b) in any other case, for at least 5 years.

Regulation 11 – Health surveillance;

1 Where it is appropriate for the protection of the health of his employees who are, or are liable to be exposed to a substance hazardous to health, the employer shall ensure that such employees are under suitable health surveillance.

2 Health surveillance shall be treated as being appropriate where:
   (a) the employee is exposed to one of the substances specified in column 1 of Schedule 5 and is engaged in a process specified in column 2 of that Schedule, unless that exposure is not significant; or
   (b) the exposure of the employee to a substance hazardous to health is such that an identifiable disease or adverse health effect may be related to the exposure, there is a reasonable likelihood that the disease or effect may occur under the particular conditions of his work and there are valid techniques for detecting indications of the disease or that effect.

3 The employer shall ensure that a health record, containing particulars approved by the HSE, in respect of each of his employees to whom paragraph 1 relates is made and maintained and that that record is kept in a suitable form for at lest 40 years from the date of the last entry made in it.

4 Where an employer who holds records in accordance with paragraph 3 ceases to trade, he shall forthwith notify the HSE in writing and offer those records to the HSE.

Paragraphs 5 to 11 deal with the following matters:

5 medical surveillance where employees are exposed to substances specified in Schedule 5;
6 prohibition by an employment medical adviser on engagement in work of employees considered to be at risk;
7 continuance of health surveillance for employees after exposure has ceased;
8 access by employees to their health records;
9 duty on employees to present themselves for health surveillance;
10 access to inspect a workplace or any record kept by employment medical advisers or appointed doctors;
11 review by an aggrieved employee or employer of medical suspension by an employment medical adviser or appointed doctor.
12 In this regulation.

appointed doctor means a registered medical practitioner who is appointed for the time being in writing by the HSE for the purposes of this regulation;

employment medical adviser means an employment medical adviser appointed under section 56 of the 1974 Act;

health surveillance includes biological monitoring.

Regulation 12 – Information, instruction and training for persons who may be exposed to substances hazardous to health

1 An employer who undertakes work which may expose any of his employees to substances hazardous to health shall provide that employee with such information, instruction and training as is suitable and sufficient for him to know:
   (a) the risks to health created by such exposure; and
   (b) the precautions which should be taken.
2 Without prejudice to the generality of paragraph 1, the information provided under that paragraph shall include
   (a) information on the results of any monitoring of exposure at the workplace in accordance with regulation 10 and, in particular, in the case of a substance hazardous to health specified in Schedule 1, the employee or his representatives shall be informed forthwith if the results of such monitoring shows that the MEL has been exceeded; and
   (b) information on the collective results of any health surveillance undertaken in a form calculated to prevent it from being identified as relating to a particular person.
3 Every employer shall ensure that any person (whether or not his employee) who carries out work in connection with the employer's

duties under these regulations has the necessary information, instruction and training.

Regulation 13 – Provisions relating to certain fumigations

1 This regulation shall apply to fumigations in which the fumigant used or intended to be used is **hydrogen cyanide, ethylene oxide, phosphine or methyl bromide**, except that this regulation shall not apply to fumigations using the fumigant specified in column 1 of Schedule 6 when the nature of the fumigation is that specified in the corresponding entry in column 2 of that Schedule.

2 An employer shall not undertake any fumigation to which this regulation applies unless he has:
  (a) notified the persons specified in Part I of Schedule 7 of his intention to undertake the fumigation; and
  (b) provided to those persons the information specified in Part II of that Schedule,

  at least 24 hours in advance, or such shorter time in advance, as the person required to be notified may agree.

3 An employer who undertakes a fumigation to which this regulation applies shall ensure that, before the fumigant is released, suitable warning notices have been affixed at all points of reasonable access to the premises or to those parts of the premises in which the fumigation is to be carried out and that after the fumigation has been completed, and the premises are safe to enter, the warning notices are removed.

Regulation 16 – Defence under the regulations

In any proceedings for an offence consisting of a contravention of these regulations it shall be a defence for any person to prove that he took **all reasonable precautions** and exercised **all due diligence** to avoid the commission of that offence.

Note

To rely on this defence, the employer must establish that, on the balance of probabilities, he has taken all precautions that were reasonable and exercised all due to diligence to ensure that these precautions were implemented in order to avoid such a contravention. It is unlikely that an employer could rely on a regulation 16 defence if:

(a) precautions were available which had not been taken; or
(b) that he had not provided sufficient information, instruction and training, together with adequate supervision, to ensure that the precautions were effective.

Thus a stated policy on the use of substances hazardous to health, company code of practice or other form of instructions to staff is insufficient without evidence of such staff being provided with the appropriate information, instruction, training and supervision.

Schedules to the COSHH Regulations

1 List of substances assigned maximum exposure limits
2 Prohibition of certain substances hazardous to health for certain purposes
3 Frequency of thorough examination and test of local exhaust ventilation plant used in certain processes
4 Specific substances and processes for which monitoring is required
5 Medical surveillance
6 Fumigations excepted from regulation 13
7 Notification of certain fumigations
8 Other substances and processes to which the definition of 'carcinogen' relates
9 Special provision relating to biological agents

Electricity at Work Regulations 1989

These regulations are made under the HSWA and apply to **all** work associated with electricity. The purpose of the regulations is to require precautions to be taken against the risk of death or personal injury from electricity in work activities.

The regulations impose duties on persons ('duty holders') in respect of systems, electrical equipment and conductors and in respect of work activities on or near electrical equipment. They replace the 1908 and 1944 regulations and extend to all premises and not just those defined as factories under the Factories Act 1961.

These regulations establish general principles for electrical safety rather than state detailed requirements. Further detailed advice is given in the supporting Memorandum of Guidance and other authoritative documents such as the IEE Wiring Regulations, British and European Standards and HSE publications.

Because the regulations state principles of electrical safety in a form which may be applied to any work activity having a bearing on electrical safety, they apply to all electrical equipment and systems wherever manufactured, purchased, installed or taken into use, even if this pre-dates them. If, however, electrical equipment does pre-date the regulations, it does not of itself mean that the continued use of such equipment is prohibited. The equipment may continue to be used provided the requirements of regulations can continue to be satisfied, in other words, the equipment need only be replaced if it becomes unsafe or there is a risk of injury.

Duties in some of the regulations are subject to the qualifying term 'reasonably practicable'. Where qualifying terms are absent, the requirements are said to be **absolute** and must be met regardless of cost or any other consideration.

Regulation 29 provides a defence for a duty holder who can establish that he took **all reasonable steps and exercised all due diligence** to avoid committing an offence under certain regulations.

Regulation 2 – Interpretation

This regulation incorporates a number of important definitions:

Circuit conductor means any conductor in a system which is intended to carry electric current in normal condition, or to be energised in normal conditions, and includes a combined neutral and earth conductor, but does not include a conductor provided solely to perform a protective function by connection to earth or other reference point.

Conductor means a conductor of electrical energy.

Danger means risk of injury.

Electrical equipment includes anything used, intended to be used or installed for use, to generate, provide, transmit, transform, rectify, convert, conduct, distribute, control, store, measure or use electrical energy.

Injury means death or personal injury from electric shock, electric burn, electrical explosion or arcing, or from fire or explosion initiated by electrical energy, where any such death or injury is associated with the generation, provision, transmission, rectification, conversion, conduction, distribution, control, storage, measurement or use of electrical energy.

System means any electrical system in which all the electrical equipment is, or may be, electrically connected to a common source of electrical energy, and includes such source and such equipment.

Regulation 3 – Persons on whom duties are imposed

Equal levels of duty are imposed on four classes of person (duty holders) i.e. employers, the self-employed, the manager of a mine or quarry, and employees.

Part II – General

Regulation 4 – Systems, work activities and protective equipment

All systems shall at all times be of such construction as to prevent, so far as is reasonably practicable, danger.

As may be necessary to prevent danger, all systems shall be maintained so as to prevent, so far as is reasonably practicable, such danger.

Every work activity, including operation, use and maintenance of a system and work near a system, shall be carried out in such a manner as not to give rise, so far as is reasonably practicable, to danger.

Any equipment provided under these regulations for the purpose of protecting persons at work on or near electrical equipment shall be suitable for the use for which it is provided, be maintained in a condition suitable for that use, and be properly used.

Regulation 5 – Strength and capability of electrical equipment

This regulation places an absolute duty to ensure that the strength and capability of electrical equipment in use are not exceeded in such a way as to give rise to danger.

The defence provided by regulation 29 is available.

Regulation 6 – Adverse or hazardous environments

Full account must be taken of any reasonably foreseeable adverse or hazardous environmental conditions that equipment may be subjected to i.e. equipment intended for outdoor use must be selected on this basis.

The adverse or hazardous environments are:

(a) mechanical danger;
(b) the effects of the weather, natural hazards, temperature or pressure;
(c) the effects of wet, dirty, dusty or corrosive conditions; or
(d) any flammable or explosive substance including dusts, vapours or gases.

Regulation 7 – Insulation, protection and placing of conductors

Regulation 7 requires that all conductors in a system which may give rise to danger (e.g. electric shock) shall either:

(a) be suitably covered with insulating material and as necessary protected so as to prevent, so far as is reasonably practicable, danger; or
(b) have such precautions taken, including their being suitably placed, as will prevent, so far as is reasonably practicable, danger.

Regulation 8 – Earthing or other suitable precautions

There is an absolute duty under the regulations to safeguard against the risk of electric shock caused by indirect contact i.e. contact with metal normally at earth potential that has become 'live' because of a fault.

The main emphasis of this regulation is on the provision of earthing. However, it is also recognised that other techniques may be employed to achieve freedom from danger e.g. double insulation, connection to a common voltage reference point on the system, equipotential bonding, use of safe voltages, earth-free non-conducting environments, current energy limitation, separated or isolated systems.

The defence provided by regulation 29 is available.

Regulation 9 – Integrity of referenced conductors

Again this regulation places an absolute duty subject to the defence provided by regulation 29. It seeks to preserve the integrity of 'referenced conductors', these being the supply system neutral conductors which are required to be connected back to earth on the low voltage side of the distribution transformer.

The object of the regulation is to prevent referenced circuit conductors, which should be at or about the same potential as the reference point, from reaching significantly different potentials thereby giving rise to possible danger.

The defence provided by regulation 29 is available.

Regulation 10 – Connections

Regulation 10 places an absolute requirement that all joints and connections in a system be mechanically and electrically suitable for use. This requirement applies whether the installation be permanent or temporary.

The defence provided by regulation 29 is available.

Regulation 11 – Means for protecting from excess of current

This regulation requires the provision of efficient means, suitably located, to protect from excess of current every part of a system as may be necessary to prevent danger.

This is an absolute requirement and the defence provided by regulation 29 is available.

Regulation 12 – Means for cutting off the supply and for isolation

Where necessary, to prevent danger, suitable means (including, where appropriate, methods of identifying circuits) shall be available for:

(a) cutting off the supply of electrical energy to any electrical equipment; and
(b) the isolation of any electrical equipment.

'Isolation' means the disconnection and separation of the electrical equipment from every source of electrical energy in such a way that this disconnection and separation is secure.

The requirements relating to isolation and separation do not apply to electrical equipment which is itself a source of electrical energy but, in such a case as is necessary, precautions shall be taken to prevent, so far as is reasonably practicable, danger.

Regulation 13 – Precautions for work on equipment made dead

Adequate precautions must be taken to prevent electrical equipment, which has been made dead in order to prevent danger while work is carried out on or near that equipment, from becoming electrically charged during that work if danger may thereby arise.

Mention is made in the Memorandum of Guidance on the need for formalisation of the isolation arrangements by the use of permit to work procedures.

Regulation 14 – Work on or near live conductors

This regulation imposes an absolute duty not to carry out work on or near any live conductor (other than one suitably covered with insulating material so as to prevent danger) such that danger may arise unless:

(a) it is unreasonable in all the circumstances for it to be dead;
(b) it is reasonable in all the circumstances for the person concerned to be at work on or near it while it is live; and
(c) suitable precautions (including where necessary the provision of suitable protective equipment) are taken to prevent injury.

A written policy statement may be needed which establishes the need for live working to be permitted.

The defence provided by regulation 29 is available.

It should be noted that regulation 28 of the 1908 Regulations imposed a duty that no person work unaccompanied when undertaking electrical work where danger may exist e.g. from live conductors. The new regulations do not automatically require accompaniment for work of this nature but this matter must be considered in the general arrangements for achieving a safe system of work.

Regulation 15 – Working space, access and lighting

There is a general requirement under this regulation for the provision of adequate working space, adequate means of access and adequate lighting at all electrical equipment on which or near which work is being done in circumstances which may give rise to danger.

Regulation 16 – Persons to be competent to prevent danger and injury

This regulation clearly states that no person shall be engaged in any work activity where technical knowledge or experience is necessary to prevent danger or, where appropriate, injury, unless he possesses such knowledge or experience, or is under such degree of supervision as may be appropriate having regard to the nature of the work.

The need, therefore, to identify 'competent persons' for certain classes of electrical work is clear. Employers must also relate the level of competence to the degree of supervision necessary.

Part III – Regulations applying to mines only

This Part deals with specific regulations applying only to mines.

Part IV – Miscellaneous and general

Regulation 29 – Defence

In any proceedings for an offence consisting of a contravention of regulations 4(4), 5, 8, 9, 10, 11, 12, 13, 14, 15, 16 or 25, it shall be a defence for any person to prove that he **took all reasonable precautions and exercised all due diligence** to avoid the commission of that offence.

Noise at Work Regulations 1989

These regulations came into operation on 1 January 1990 and are accompanied by a number of Noise Guides issued by the HSE. They give effect in Great Britain to provisions of Council Directive 86/188/EEC on the protection of workers from the risks related to exposure to noise. The regulations bring in the concept of 'daily personal noise exposure' and 'action levels'.

Daily personal noise exposure relates to the level of daily personal noise exposure of an employee ascertained in accordance with Part 1 of the

Schedule to the regulations, but taking no account of the effect of any personal ear protector used.

Exposed means exposed while at work, and **exposure** shall be construed accordingly.

The **first action level** means a daily personal noise exposure of 85 dB(A).

The **peak action level** means a peak sound pressure of 200 pascals.

The **second action level** means a daily personal noise exposure of 90 dB(A).

Regulation 4 – Assessment of exposure

Under regulation 4 every employer shall, when any of his employees is likely to be exposed to the first action level or above or to the peak action level or above, ensure that a competent person makes a noise assessment which is adequate for the purposes of:

(a) identifying which of his employees are so exposed; and
(b) providing him with such information with regard to the noise to which those employees may be exposed as will facilitate compliance with his duties under regulations 7, 8, 9 and 11.

The above noise assessment shall be reviewed when:

(a) there is reason to suspect that the assessment is no longer valid; or
(b) there has been a significant change in the work to which the assessment relates.

Where as a result of the review, changes in the assessment are required, those changes shall be made.

Regulation 5 – Assessment records

Following any noise assessment, the employer shall ensure that an adequate record of the assessment and of any review thereof carried out, is kept until a further noise assessment is made.

Regulation 6 – Reduction of risk of hearing damage

This regulation places a general duty on every employer to reduce the risk of damage to the hearing of his employees from exposure to noise to the lowest level reasonably practicable.

Regulation 7 – Reduction of noise exposure

Every employer shall, when any of his employees is likely to be exposed to the second action level or above or to the peak action level or above, reduce, so far as is reasonably practicable (other than by the provision of ear protectors), the exposure to noise of that employee.

Regulation 8 – Ear protection

Whilst the emphasis under the regulations is on noise control, the regulations do recognise the need for ear protection. Under regulation 8 every employer shall ensure, so far as is practicable, that when any of his employees is likely to be exposed to the **first action level** or above in circumstances where the daily personal noise exposure of that employee is likely to be less than 90 dB(A), that employee is provided, at his request, with suitable and sufficient personal ear protectors.

Furthermore, when any of his employees is likely to be exposed to the **second action level** or above or to the **peak action level** or above, every employer shall ensure, so far as is practicable, that that employee is provided with suitable ear protectors which, when properly worn, can reasonably be expected to keep the risk of damage to that employee's hearing to below that arising from exposure to the second action level or, as the case may be, to the peak action level.

Regulation 8 – Ear protection zones

Regulation 8 is concerned with the demarcation and identification of ear protection zones i.e. areas where employees must wear ear protection.

Thus every employer shall, in respect of any premises under his control, ensure, so far as is reasonably practicable, that:

(a) each ear protection zone is demarcated and identified by means of the sign specified in paragraph A.3.3 of Appendix A to Part 1 of BS 5878, which sign shall include such text as indicates:
  (i) that it is an ear protection zone; and
  (ii) the need for his employees to wear personal ear protectors while in any such zone; and
(b) none of his employees enters any such zone unless that employee is wearing personal ear protectors.

In this regulation, **ear protection zone** means any part of the premises referred to above where any employee is likely to be exposed to the second action level or above or to the peak action level or above, and **Part 1 of BS 5378** has the same meaning as in regulation 2(1) of the Safety Signs Regulations 1980.

Regulation 10 – Maintenance and use of equipment

Every **employer** shall:

(a) ensure, so far as is practicable, that anything provided by him to or for benefit of an employee in compliance with his duties under these regulations (other than personal ear protectors provided pursuant to regulation 8) is fully and properly used; and
(b) ensure, so far as is practicable, that anything provided by him in compliance with his duties under these regulations is maintained in **an efficient state, in efficient working order and in good repair**.

Every **employee** shall, so far as is practicable, fully and properly use personal ear protectors when they are provided by his employer pursuant to regulation 7 and any other protective measures provided by his employer in compliance with his duties under these regulations; and if the employee discovers any defect therein, he shall report it forthwith to his employer.

Regulation 11 – Provision of information to employees

Every employer shall, in respect of any premises under his control, provide each of his employees who is likely to be exposed to the first action level or above or to the peak action level or above, with adequate information, instruction and training on:

(a) the risk of damage to that employee's hearing that such exposure may cause;
(b) what steps that employee can take to minimise that risk;
(c) the steps that the employee **must** take in order to obtain the personal ear protectors referred to in regulation 8; and
(d) that employee's obligations under these regulations.

Regulation 12 – Modification of duties of manufacturers etc. of articles for use at work and articles of fairground equipment

The duties under section 6 of HSWA on the part of manufacturers, designers, etc. were modified to include a duty to ensure that, where any such article is likely to cause any employee to be exposed to the first action level or above or to the peak action level or above, adequate information is provided concerning the noise likely to be generated by that article.

Personal Protective Equipment at Work Regulations 1992

These regulations:

(a) amend certain regulations made under the HSWA which deal with personal protective equipment (PPE), so that they fully implement the European Directive in circumstances where they apply;
(b) cover all aspects of the **provision, maintenance and use** of PPE at work in other circumstances; and
(c) revoke and replace almost all pre-HSWA and some post-HSWA legislation which deals with PPE.

Specific requirements of current regulations dealing with PPE, namely the Control of Lead at Work Regulations 1980, the Ionising Radiations Regulations 1985, the Control of Asbestos at Work Regulations 1987, the Control of Substances Hazardous to Health (COSHH) Regulations 1994, the Noise at Work Regulations 1989 and the Construction (Head Protection) Regulations 1989, take precedence over the more general requirements of the Personal Protective Equipment at Work Regulations 1992.

Regulation 2 – Interpretation

Personal protective equipment means all equipment (including clothing affording protection against the weather) which is intended to be worn or held by a person at work and which protects him against one or more risks to his health and safety, and any addition or accessory designed to meet this objective.

Regulation 3 – Disapplication of these regulations

1 These regulations shall not apply to or be in relation to the master or crew of a sea-going ship or to the employer of such persons in respect of normal shipboard activities of a ship's crew under the direction of a master.
2 Regulations 4 to 12 shall not apply in respect of personal protective equipment which is:
   (a) ordinary working clothes and uniforms which do not specifically protect the health and safety of the wearer;
   (b) an offensive weapon within the meaning of section 1(4) of the Prevention of Crime Act 1953 used as self-defence or deterrent equipment;
   (c) portable devices for detecting and signalling risks and nuisances;
   (d) personal protective equipment used for protection while travelling on a road;
   (e) equipment used during the playing of competitive sports.

Regulation 4 – Provision of personal protective equipment

1 Every employer shall ensure that suitable PPE is provided to his employees who may be exposed to a risk to their health and safety while at work except where and to the extent that such risk has been adequately controlled by other means which are equally or more effective.
2 Similar provisions as above apply in the case of self-employed persons.
3 PPE shall not be suitable unless:
   (a) it is appropriate for the risk or risks involved and the conditions at the place where exposure to the risk may occur;
   (b) it takes account of ergonomic requirements and the state of health of the person or persons who may wear it;
   (c) it is capable of fitting the wearer correctly, if necessary after adjustments within the range for which it is designed;
   (d) **so far as is practicable**, it is effective to prevent or adequately control the risk or risks involved without increasing overall risk.

Regulation 5 – Compatibility of personal protective equipment

1 Every employer shall ensure that where the presence of more than one risk to health or safety makes it necessary for his employee to wear or use simultaneously more than one item of PPE, such equipment is compatible and continues to be effective against the risk or risks in question.
2 Similar provisions apply in the case of self-employed persons.

Regulation 6 – Assessment of personal protective equipment

1 Before choosing any personal protective equipment (PPE) which he is

required to provide, an employer or self-employed person shall make an assessment to determine whether the PPE he intends to provide is suitable.

2 The assessment shall comprise:
   (a) an assessment of any risk or risks which have not been avoided by other means;
   (b) the definition of the characteristics which PPE must have in order to be effective against the risks referred to above, taking into account any risks which the equipment itself may create;
   (c) comparison of the characteristics of the PPE available with the characteristics referred to in (b) above.

The assessment shall be reviewed forthwith if:

(a) there is reason to suspect that any element of the assessment is no longer valid; or
(b) there has been a significant change in the work to which the assessment relates;

and where, as a result of the review, changes in the assessment are required, these changes shall be made.

Regulation 7 – Maintenance and replacement of personal protective equipment

Every employer and every self-employed person shall ensure that any PPE provided by them is maintained, in relation to any matter which it is **reasonably foreseeable** will affect the health and safety of any person, in an efficient state, in efficient working order, in good repair and in hygienic condition.

Regulation 8 – Accommodation for personal protective equipment

Every employer and every self-employed person shall ensure that appropriate accommodation is provided for PPE when it is not being used.

Regulation 9 – Information, Instruction and training

Where an employer is required to provide PPE to an employee, the employer shall provide that employee with such information, instruction and training as is adequate and appropriate to enable the employee to know:

(a) the risk or risks which the PPE will avoid or limit;
(b) the purpose for which and the manner in which the PPE is to be used; and
(c) any action to be taken by the employee to ensure that the PPE remains in an efficient state, in efficient working order, in good repair and in hygienic condition.

Regulation 10 – Use of personal protective equipment

1 Every employer who provides any PPE shall take all reasonable steps to ensure that it is properly used.

2 Every employee and self-employed person who has been provided with PPE shall:
   (a) make full and proper use of the PPE; and
   (b) take all reasonable steps to ensure it is returned to the accommodation provided for it after use.

Regulation 11 – Reporting loss or defect
Every employee who has been provided with PPE by his employer shall forthwith report to his employer any loss of or obvious defect in that PPE.

Guidance on the regulations
Detailed HSE guidance is provided on the requirements of the regulations.

Health and Safety (Display Screen Equipment) Regulations 1992

These regulations came into operation on 1 January 1993 and should be read in conjunction with the Management of Health and Safety at Work Regulations 1992. Under the regulations, employers have a duty to ensure that newly established workstations i.e. those established after 1 January 1993, and those which have been subject to modification, meet the requirements laid down in the Schedule immediately. Existing workstations i.e. those in operation prior to 1 January 1993, must comply with the Schedule not later than 31 December 1996. It is important when considering the implementation of these regulations to appreciate four definitions shown in regulation 1:

Display screen equipment means an alphanumeric or graphic display screen, regardless of the display process involved.

Operator means a self-employed person who **habitually** uses display screen equipment as a **significant** part of his normal work.

User means an employee who **habitually** uses display screen equipment as a **significant** part of his normal work.

Workstation means an assembly comprising:

(a) display screen equipment (DSE) (whether provided with software determining the interface between the equipment and its operator or user, a keyboard or any other input device);
(b) any optional accessories to the DSE;
(c) any disk drive, modem, printer, document holder, work chair, work desk, work surface or other item peripheral to the DSE; and
(d) the immediate environment around the DSE.

The terms 'habitual' and 'significant' in the definition of both 'user' and 'operator' are important in that many people who use DSE as a feature of their work activities may not necessarily be users as defined. Employers must decide which of their employees are 'users' by referring to the Guidance which accompanies the regulations.

The regulations do NOT apply to or in relation to:

(a) drivers' cabs or control cabs;
(b) DSE onboard a means of transport;
(c) DSE mainly intended for public operation;
(d) portable systems not in prolonged use;
(e) calculators, cash registers or any equipment having a small data or measurement display required for direct use of the equipment; or
(f) window typewriters.

Regulation 2 – Analysis of workstations

Regulation 2 is the principal requirement. Employers shall perform a suitable and sufficient analysis of those workstations which:

(a) (regardless or who has provided them) are used for the purposes of his undertaking by users; and
(b) have been provided by him and are used for the purposes of his undertaking by operators;

for the purpose of assessing the health and safety risks to which those persons are exposed in consequence of that use.

The assessment must be reviewed when it is no longer valid or there has been a significant change in the matters to which it relates. The employer shall then reduce the risks identified to the lowest extent reasonably practicable.

Regulation 4 – Daily work routines

This regulation requires that every employer shall plan the activities of users at work in his undertaking that their daily work routine is periodically interrupted by such breaks or changes of activity as reduce their workload at that equipment. This is a matter that needs addressing for certain groups of users who spend a substantial part of their time in front of a DSE. Information on this aspect should be derived from the risk assessment carried out.

Regulation 5 – Eye and eyesight tests

Employers are required to make provision on request for eye and eyesight tests for existing users and people who become users, such tests to be undertaken by a competent person. A competent person in this case, according to the Guidance, is a registered ophthalmic optician or a registered medical practitioner with suitable qualifications. (It is important to distinguish here between an 'eye and eyesight test' and 'vision screening', the latter frequently being undertaken by occupational health nurses.)

Regulation 6 – Provision of training

Employers shall ensure that existing and new users of VSE receive adequate health and safety training in the use of any workstation upon which they may be required to work. The Guidance goes into great detail on the range of training required according to specific circumstances.

Regulation 7 – Provision of information

Regulation 7 states that every employer shall ensure that operators and users are provided with adequate information about all aspects of health and safety relating to their workstations, and such measures taken by him in compliance with his duties under regulations 2 and 3. Similarly, information must be given to users about measures taken to comply with regulations 5 and 6 as relate to them and their work.

DSE risk analysis

Risk analysis involves a consideration of the requirements laid down in the Schedule to the regulations relating to the equipment, the environment and the interface between the computer and operator/user.

The Chemicals (Hazard Information and Packaging for Supply) Regulations 1994

These regulations (known as the CHIP 2 Regulations) came into force on 31 January 1995. They revoke the former Chemicals (Hazard Information and Packaging) Regulations (CHIP 1) 1993. Moreover, the 'carriage' requirements, formerly detailed in CHIP 1, were transferred to the Carriage of Goods by Road and Rail (Classification, Packaging and Labelling) Regulations 1994.

The regulations cover, in particular, the requirements for the classification, packaging and labelling of both *substances* and *preparations* which are *dangerous for supply* and the information to be provided in *safety data sheets*. Specific requirements for child-resistant fastenings, tactile warning devices and notification arrangements for certain preparations dangerous for supply to the poisons advisory centre are also dealt with in the regulations. The Schedules to the regulations deal with the actual classifications, the indications of danger and symbols necessary and the headings to be incorporated in safety data sheets.

A number of important definitions are dealt with below:

Category of danger means in relation to a substance or preparation dangerous for supply, one of the categories of danger specified in column 1 of Part I of Schedule 1.

Indication of danger means, in relation to a substance or preparation dangerous for supply, one or more of the indications of danger referred to in column 1 of Schedule 2 and:

(a) in the case of a substance dangerous for supply listed in part I of the Approved Supply List, it is one or more indications of danger for that substance specified by a symbol–letter in column 3 of Part V of that list; or

(b) in the case of a substance dangerous for supply *not* so listed or a preparation dangerous for supply, it is one or more indications of danger determined in accordance with the classification of that substance or preparation under regulation 5 and the Approved Classification and Labelling Guide.

Note

The Approved Supply List indicates the specific classification and labelling requirements for some 1400 substances that are already agreed with the EU. The Approved Guide to the Classification and Labelling of Substances and Preparations Dangerous for Supply must be used by the supplier in the self-classification and labelling of substances and preparations that are not listed in the Approved Supply List.

Package means, in relation to a substance or preparation dangerous for supply, the package in which the substance or preparation is supplied and which is liable to be individually handled during the course of the supply and includes the receptacle containing the substance or preparation and any other packaging associated with it and any pallet or other device which enables more than one receptacle containing a substance or preparation dangerous for supply to be handled as a unit, but does not include:

(a) a freight container (other than a tank container), a skip, a vehicle or other article of transport equipment; or
(b) in the case of supply by way of retail sale, any wrapping such as a paper or plastic bag into which the package is placed when it is presented to the purchaser.

9

Packaging means, in relation to a substance or preparation dangerous for supply, as the context may require, the receptacle, or any components, materials or wrappings associated with the receptacle for the purpose of enabling it to perform its containment function or both.

Preparation means mixtures or solutions of two or more substances.

Preparations dangerous for supply means a preparation which is in one or more categories of danger specified in column 1 of Schedule 1.

Substance dangerous for supply means:

(a) a substance listed in Part I of the Approved Supply List; or
(b) any other substance which is in one or more of the categories of danger specified in column 1 of Schedule 1.

Risk phrase means, in relation to a substance or preparation dangerous for supply, a phrase listed in Part III of the Approved Supply List and in these regulations specific risk phrases may be designated by the letter 'R' followed by a distinguishing number or combination of numbers but the risk phrase shall be quoted in full on any label or safety data sheet in which the risk phrase is required to be shown.

Safety phrase means, in relation to a substance or preparation dangerous for supply, a phrase listed in Part IV of the Approved Supply List and in these regulations specific safety phrases may be designated by the letter 'S' followed by a distinguishing number or combination of numbers, but the safety phrase shall be quoted in full on any label or safety data sheet in which the safety phrase is required to be shown.

Supplier means a person who supplied a substance or preparation dangerous for supply, and in the case of a substance which is imported (whether or not from a member state) includes the importer established in Great Britain of that substance or preparation.

Supply in relation to a substance or preparation:

(a) means, subject to paragraphs (b) or (c) below, supply of that substance or preparation, whether as principal or agent for another, in the course of or for use at work, by way of:
 (i) sale or offer for sale;
 (ii) commercial sample; or
 (iii) transfer from a factory, warehouse or other place of work and its curtilage to another place of work, whether or not in the same ownership;

(b) for the purposes of sub-paragraphs (a) and (b) of regulation 16(2) (HSE as enforcement agency), except in relation to regulations 7 (advertisements) and 12 (child-resistant fastenings and warning devices), in any case for which by virtue of those sub-paragraphs the enforcing authority for these regulations is the Royal Pharmaceutical Society or the local weights and measures authority, has the meaning assigned to it by section 46 of the Consumer Protection Act 1987 and also includes *offer* to supply and *expose* for supply; or

(c) in relation to regulations 7 and 12, shall have the meaning assigned to it by regulation 7(2) and 12(2) respectively.

Application of the regulations

These regulations apply to any substance or preparation which is dangerous for supply *except*

(a) radioactive substances or preparations;
(b) animal feeds;
(c) cosmetic products;
(d) medicines and medicinal products;
(e) controlled drugs;
(f) substances or preparations which contain disease-producing micro-organisms;
(g) substances or preparations taken as samples under any enactment;
(h) munitions, which produce explosion or pyrotechnic effect;
(i) foods;
(j) a substance or preparation which is under customs control;
(k) a substance which is intended for export to a country which is not a member state;
(l) pesticides;
(m) a substance or preparation transferred with a factory, warehouse or other place of work;
(n) a substance to which regulation 7 of the Notification of New Substances Regulations 1993 applies; or
(o) substances, preparations or mixtures in the form of wastes.

Regulation 5 – Classification of substances and preparations

Under regulation 5 a supplier shall not supply a substance or preparation dangerous for supply unless it has been classified in accordance with criteria laid down in paragraphs 2–6, which cover substances listed in the Approved Supply List, new substances, other substances dangerous for supply, preparations to which the regulations apply, and pesticides.

Regulation 6 – Safety data sheets

Regulation 6 places an absolute duty, with certain exceptions, on the supplier to provide the recipient with a safety data sheet containing information under the headings specified in Schedule 5 to enable the recipient of the substance or preparation to take the necessary measures relating to the protection of health and safety at work and relating to the protection of the environment. A safety data sheet must show its date of first publication or latest revision.

Schedule 5 lists the following obligatory headings which must be contained in a safety data sheet:

1 Identification of the substance/preparation
2 Composition/information on ingredients
3 Hazards identification
4 First aid measures
5 Fire-fighting measures
6 Accidental release measures
7 Handling and storage
8 Exposure controls/personal protection
9 Physical and chemical properties
10 Stability and reactivity
11 Toxicological information
12 Ecological information
13 Disposal considerations
14 Transport information
15 Regulatory information
16 Other information.

Further information is provided in the Approved Code of Practice 'Safety Data Sheets for Substances and Preparations Dangerous for Supply'.

Regulation 7 – Advertisements

Regulation 7 places an absolute duty on a person who supplies or offers a substance dangerous for supply to ensure mention is made in the advertisement of the hazard or hazards presented by the substance.

Regulation 8 – Packaging requirements

Generally, a package must be suitable for its purpose unless:

(a) it is designed, constructed, maintained and closed so as to prevent escape of the contents;

(b) is made of materials which are neither liable to be adversely affected by the substance nor liable, in conjunction with the substance, to form another substance which is itself a risk to health or safety; and
(c) where fitted with a replaceable close, the closure is designed to permit repeated re-closure without escape of the contents (regulation 8).

Regulation 9 – Labelling

Regulation 9 lays down the labelling criteria for both substances and preparations dangerous for supply, in terms of the information to be provided on labels, as follows:

(a) Substances dangerous for supply:
  the name, full address and telephone number of the supplier (as defined);
  the name of the substance;
  the following particulars:
  (i) the indication or indications of danger and the corresponding symbol or symbols;
  (ii) the risk phrases;
  (iii) the safety phrases;
  (iv) the EEC number (if any) and, in the case of a substance dangerous for supply which is listed in Part I of the Approved Supply List, the words 'EEC label';
(b) Preparations dangerous for supply:
  the name, full address and telephone number of the supplier;
  the trade name or other designation of the preparation;
  the following particulars:
  (i) identification of the constituents of the preparation which result in the preparation being classified as dangerous for supply;
  (ii) the indication or indications of danger and the corresponding symbol or symbols;
  (iii) the risk phrases;
  (iv) the safety phrases;
  (v) in the case of a pesticide, the modified information specified in paragraph 5 of Part I of Schedule 6; and
  (vi) in the case of a preparation intended for sale to the general public, the nominal quantity (nominal mass or nominal volume); and
  where required by paragraph 5(5) of Part I of Schedule 3, the words specified in that paragraph.

Regulation 10 deals with particular labelling requirements for certain preparations to which Part II of Schedule 6 applies.

Regulation 11 – Methods of marking or labelling packages

Any package which is required to be labelled in accordance with regulations

9 and 10 must comply with the following requirements under regulation 11.

(a) the particulars required to be on the label must be clearly and indelibly marked on that part of the package reserved for that purpose;
(b) the label shall be securely fixed;
(c) the colour and nature of the marking shall be such that the symbol and wording stand out from the background so as to be readily recognisable and easily read;
(d) the package to be labelled so that the particulars can be read horizontally;
(e) the dimensions of the label to be related to the capacity of the package as specified in regulation 11(5);
(f) any symbol required to be shown in accordance with regulation 9(2)(c)(i) or 9(3)(c)(ii) and specified in column 3 of Schedule 2 shall be printed in black on an orange–yellow background;
(g) where the package is of an awkward shape or so small that it is unsuitable to attach a label that complies with the above, the label shall be so attached in some other appropriate manner; and
(h) the particulars shall be in English except that where supplied to a recipient in a member state, the label shall be in the official language of the state.

Regulation 12 – Child-resistant fastenings and tactile warning devices
Regulation 12 lays down specific provisions based on British and International Standards which are further described in Schedule 7.

Regulation 13 – Retention of classification data
Under regulation 13 a person who classifies substances and preparations dangerous for supply shall keep a record of the information used for the purpose of classifying for at least 3 years after the date on which the substance or preparation was supplied to him, making such record available to the enforcing authority on request.

Regulation 14 – Poisons Advisory Centre
Regulation 14 places a duty on suppliers of certain preparations, classified as dangerous for supply on the basis of one or more their health effects, to notify the Poisons Advisory Centre of the information required to be in the safety data sheet.

Regulation 16 – Enforcement arrangements, civil liability and the defence available
Regulation 16 deals with the following matters:

(a) subject to regulation 16(2) the provisions of HSWA, which relate to the approval of ACOPs and their use in criminal proceedings, to enforcement and offences shall apply; and
(b) a breach of duty imposed shall confer a right of action in civil proceedings.

The enforcing authority is the HSE, except that.

(a) where a substance or preparation dangerous for supply is supplied in or from premises which are registered under section 75 of the Medicines Act 1968 the enforcing authority shall be the Royal Pharmaceutical Society;

(b) where supplied otherwise:
   (i) in or from any shop, mobile vehicle, market stall or other retail outlet; or
   (ii) otherwise to members of the public, including by way of free sample, prize or mail order,

   the enforcing authority shall be the local weights and measures authority; and

(c) for regulations 7 and 12 the enforcing authority shall be the local weights and measures authority.

The maximum period of imprisonment on summary conviction shall be 3 months.

In any proceedings for an offence under these regulations, it shall be a defence for any person to prove that he took **all reasonable precautions and exercised all due diligence** to avoid the commission of that offence.

SCHEDULE I

CLASSIFICATION OF SUBSTANCES AND PREPARATIONS DANGEROUS FOR SUPPLY

PART I
CATEGORIES OF DANGER

Column 1 Category of danger	Column 2 Property (see Note 1)	Column 3 Symbol–letter
PHYSICO-CHEMICAL PROPERTIES		
Explosive	Solid, liquid, pasty or gelatinous substances and preparations which may react exothermically without atmospheric oxygen thereby quickly evolving gases, and which under defined test conditions detonate, quickly deflagrate or upon heating explode when partially confined	E
Oxidising	Substances and preparations which give rise to an exothermic reaction in contact with other substances, particularly flammable substances	0
Extremely flammable	Liquid substances and preparations having an extremely low flash point and a low boiling point and gaseous substances and preparations which are flammable in contact with air at ambient temperature and pressure	F+
Highly flammable	The following substances and preparations, namely: (a) substances and preparations which may become	

Column 1 Category of danger	Column 2 Property (see Note 1)	Column 3 Symbol–letter
	hot and finally catch fire in contact with air at ambient temperature without any application of energy; (b) solid substances and preparations which may readily catch fire after brief contact with a source of ignition and which continue to burn or to be consumed after removal of the source of ignition; (c) liquid substances and preparations having a very low flash point; (d) substances and preparations which, in contact with water or damp air, evolve highly flammable gases in dangerous quantities (see Note 2).	F
Flammable	Liquid substances and preparations having a low flash point.	None
HEALTH EFFECTS		
Very toxic	Substances and preparations which in **very low quantities** can cause death or acute or chronic damage to health when inhaled, swallowed or absorbed via the skin.	T+
Toxic	Substances and preparations which in **low quantities** can cause death or acute or chronic damage to health when inhaled, swallowed or absorbed via the skin.	T
Harmful	Substances and preparations which cause death or acute or chronic damage to health when inhaled, swallowed or absorbed via the skin.	Xn
Corrosive	Substances and preparations which may, on contact, with living tissues, **destroy** them.	C
Irritant	Non-corrosive substances and preparations which through immediate, prolonged or repeated contact with the skin or mucous membrane, may cause **inflammation**.	Xi
Sensitising	Substances and preparations which, if they are inhaled or if they penetrate the skin, are capable of eliciting a reaction by **hypersensitisation** such that on further exposure to the substance or preparation, characteristic adverse effects are produced.	
Sensitising by inhalation		Xn
Sensitising by skin contact		Xi
Carcinogenic (See Note 3)	Substances and preparations which, if they are inhaled or ingested or if they penetrate the skin, may induce **cancer** or increase its incidence.	

Column 1 Category of danger	Column 2 Property (see Note 1)	Column 3 Symbol–letter
Category 1 Category 2 Category 3		T T Xn
Mutagenic (See Note 3)	Substances and preparations which, if they are inhaled or ingested or if they penetrate the skin, may induce **heritable genetic defects** or increase their incidence.	
Category 1 Category 2 Category 3		T T Xn
Toxic for repro–duction (See Note 3)	Substances and preparations which, if they are inhaled or ingested or if they penetrate the skin, may produce or increase the incidence of **non-heritable adverse effects** in the progeny and/or an impairment of male or female reproductive functions or capacity.	
Category 1 Category 2 Category 3		T T Xn
Dangerous for the environment (See Note 4)	Substances which, were they to enter into the environment, would present or might present an immediate or delayed danger for one or more components of the environment	

Notes

1 As further described in the Approved Classification and Labelling Guide.

2 Preparations packed in aerosol dispensers shall be classified as flammable in accordance with the additional criteria set out in Part II of this Schedule.

3 The categories are specified in the Approved Classification and Labelling Guide.

4 (a) In certain cases specified in the Approved Supply List and in the Approved Classification and Labelling Guide as dangerous for the environment do not require to be labelled with the symbol for this category of danger.

(b) This category of danger does not apply to preparations.

Bibliography and further reading

Chapter 1

Stranks, J., *The Handbook of Health and Safety Practice* (third edition) (Pitman, 1994)

International Labour Organisation, *Occupational Health Services* (ILO, Geneva, 1984)

Royal College of Nursing, *An Occupational Health Service* (RCN, London, 1975)

Schilling, R.S.F., *Occupational Health Practice* (Butterworths, London, 1975)

Health and Safety Executive, *Guidelines for Occupational Health Services* (HMSO, 1982)

Health and Safety Executive, *Pre-employment Health Screening: Guidance Note MS20* (HMSO, 1982)

Health and Safety Executive, *Surveillance of People Exposed to Health Risks at Work: Guidance Note HS(G)61* (HMSO, 1990)

Health and Safety Executive, *Protecting Your Health at Work* (HSE Information Centre, Sheffield, 1993)

Chapter 2

Chemicals (Hazard Information and Packaging for Supply) Regulations 1994 (SI 1994 No. 3247) (HMSO, 1995)

Health and Safety Commission, *Approved Supply List: Information Approved for the Classification and Labelling of Substances and Preparations Dangerous for Supply: Chemicals (Hazard Information and Packaging for Supply) Regulations 1994* (HSE Books, 1995)

Health and Safety Commission, *Approved Guide to the Classification and Labelling of Substances and Preparations Dangerous for Supply: Guidance on Regulations; Chemicals (Hazard Information and Packaging for Supply) Regulations 1994* (HSE Books, 1995)

Health and Safety Commission, *Approved Code of Practice: Safety Data Sheets for Substances and Preparations Dangerous for Supply: Chemicals (Hazard Information and Packaging for Supply) Regulations 1994* (HSE Books, 1995)

Health and Safety Executive, *The Popular Guide: CHIP for Everyone* (HSE Books 1995)

Health and Safety Executive, *The Complete Idiot's Guide to CHIP* (HSE Books, 1995)

Health and Safety Executive, *Why do I Need a Safety Data Sheet?* (HSE Books, 1995)

Health and Safety Executive, *Operational Provisions of the Dangerous Substances (Conveyance by Road in Road Tankers and Tank Containers) Regulations 1981* (HMSO 1981)

Health and Safety Executive, *Transport of Dangerous Substances in Tank Containers: Guidance booklet HS(G)27* (HMSO, 1986)

Health and Safety Executive, *Approved Code of Practice: Control of Lead at Work* (HMSO, 1980)

Health and Safety Commission, Ceramics Industry Advisory Committee, *Lead: A Guide to Assessment* (HSE Information Centre, Sheffield, 1992)

Health and Safety Executive, *A Step by Step Guide to COSHH Assessment* (HMSO, 1993)

Stranks, J., *The Handbook of Health and Safety Practice* (third edition) (Pitman, 1994)

Health and Safety Executive, *Approved Code of Practice: Control of Asbestos at Work; Control of Asbestos at Work Regulations 1987* (HMSO, 1987)

Health and Safety Executive, *Approved Code of Practice: Work with Asbestos Insulation, Asbestos Coating and Asbestos Insulating Board; Control of Asbestos at Work Regulations 1987* (HMSO, 1988)

Health and Safety Executive, *Approved Code of Practice: Control of Vinyl Chloride at Work* (HMSO, 1988)

Department of Health and Social Security, *Notes on the Diagnosis of Occupational Diseases* (HMSO, 1983)

Atherley, G.R.C., *Occupational Health and Safety Concepts* (Applied Science Publishers Ltd, London, 1978)

Chapter 3

Health and Safety Executive, *Ionising Radiations Regulations 1985 (SI 1985 No1333)* (HMSO, 1985)

Health and Safety Executive, *Approved Code of Practice: Protection of Persons Against Ionising Radiation Arising from any Work Activity; Ionising Radiations Regulations 1985* (HMSO, 1985)

Health and Safety Executive, *Wear Your Film Badge* (HSE Enquiry Points, London and Sheffield)

Stranks, J., *The Handbook of Health and Safety Practice* (third edition) (Pitman, 1994)

Bilsom International, *In Defence of Hearing* (Bilsom International, Henley-on-Thames, 1992)

Bruel & Kjaer, *Measuring Sound* (Bruel & Kjaer, Naerum, Denmark, 1984)

Burns, W., *Noise and Man* (John Murray, London, 1973)

Health and Safety Executive, *100 Applications of Noise Reduction Methods* (HMSO, 1983)

Health and Safety Executive, *Noise at Work: Noise Guides Nos. 1 to 8; Noise at Work Regulations 1989* (HMSO, 1989/1990)

Health and Safety Commission, *Vibration White Finger in Foundries: Advice for Employers* (HSE Enquiry Points, London and Sheffield, 1992)

Health and Safety Executive, *Work-related Upper Limb Disorders: A Guide to Prevention: Guidance Note HS(G)60* (HMSO, 1992)

Health and Safety Executive, *Upper Limb Disorders: Assessing the Risks* (HSE Books, 1994)

Health and Safety Executive, *Memorandum of Guidance on the Electricity at Work Regulations 1989: Guidance on Regulations* (HMSO, 1989)

Health and Safety Executive, *Display Screen Equipment Work: Guidance on Regulations; Health and Safety (Display Screen Equipment) Regulations 1992* (HMSO, 1992)

Health and Safety Executive, *Working with VDUs* (HSE Enquiry Points, London and Sheffield, 1992)

The Central Computer and Telecommunications Agency and the Council of the Civil Service Unions, *Ergonomic Factors Associated with the Use of Visual Display Units* (CCTA, London, 1988)

Chapter 4

Department of Health and Social Security, *Notes on the Diagnosis of Occupational Diseases* (HMSO, 1983)

Health and Safety Executive, *Legionnaire's Disease: Guidance Note EH48* (HMSO, 1987)

Health and Safety Executive, *Approved Code of Practice: The Prevention or Control of Legionellosis (including Legionnaire's Disease)* (HMSO, 1991)

Health Education Authority, *The AIDS Test* (HEA, London, 1989)

Stranks, J., *The Handbook of Health and Safety Practice* (third edition) (Pitman, 1994)

Chapter 5

Health and Safety Executive, *Occupational Exposure Limits: Guidance Note EH40* (HMSO, 1994)

Health and Safety Commission, *Approved Code of Practice: Safety Data Sheets for Substances and Preparations Dangerous for Supply; Chemicals (Hazard Information and Packaging) Regulations 1994* (HSE Books, 1995)

Plunkett, E. R., *Handbook of Industrial Toxicology* (Heyden & Son, London, 1976)

Stranks, J., *The Handbook of Health and Safety Practice* (third edition) (Pitman, 1994)

Chapter 6

Health and Safety Executive, *Monitoring Strategies for Toxic Substances: Guidance Note EH42* (HMSO, 1989)

Stranks, J., *The Handbook of Health and Safety Practice* (third edition) (Pitman, 1994)

Bruel & Kjaer, *Measuring Sound* (Bruel & Kjaer, Naerum, Denmark, 1984)

Health and Safety Executive, *Approved Code of Practice: Protection of Persons Against Ionising Radiation Arising from any Work Activity; Ionising Radiations Regulations 1985* (HMSO, 1985)

Health and Safety Executive, *Surveillance of People Exposed to Health Risks at Work* (HMSO, 1990)

Chapter 7

Health and Safety Executive, *COSHH: A brief guide for employers* (HSE Enquiry Points, London and Sheffield, 1993)

Health and Safety Executive, *The Maintenance, Examination and Testing of Local Exhaust Ventilation: Guidance Note HS(G)54* (HMSO, 1990)

Health and Safety Executive, *An Introduction to Local Exhaust Ventilation: Guidance Note HS(G)37* (HMSO, 1987)

Stranks, J., *The Handbook of Health and Safety Practice* (third edition) (Pitman, 1994)

American Conference of Government Industrial Hygienists, *Industrial Ventilation* (19th edition) (ACGIH, Pittsburg, USA)

Health and Safety Executive, *Lighting at Work: Guidance Note HS(G)38* (HMSO, 1987)

Chapter 8

Health and Safety Executive, *Respiratory Protective Equipment: A Guide for Users: Guidance booklet HS(G)53* (HMSO, 1990)

Health and Safety Executive, *Personal Protective Equipment at Work: Guidance on Regulations; Personal Protective Equipment at Work Regulations 1992* (HMSO, 1992)

Health and Safety Executive, *Approved Code of Practice: Control of Lead at Work* (HMSO, 1985)

Health and Safety Executive, *Construction (Head Protection) Regulations 1987: Guidance on Regulations* (HMSO, 1990)

Health and Safety Executive, *Approved Code of Practice and Guidance: Safety in Docks; Docks Regulations 1988* (HMSO, 1988)

Health and Safety Executive, *Protective Clothing and Footwear in the Construction Industry* (HMSO, 1991)

Health and Safety Executive, *Safety with Chainsaws* (HMSO, 1990)

Health and Safety Executive, *Approved Code of Practice: Work with Asbestos Insulation, Asbestos Coating and Asbestos Insulating Board* (HMSO, 1988)

British Standards Institute, *BS 4275: Recommendations for the Selection, Use and Maintenance of Personal Protective Equipment* (BSI, 1988)

Health and Safety Executive, *Noise Guide No.5: Types and Selection of Personal Ear Protectors* (HMSO, 1990)

Health and Safety Executive, *Respiratory Protective Equipment: Legislative Requirements and Lists of HSE Approved Standards and Type Approved Equipment* (HMSO, 1992)

Chapter 9

Secretary of State for Employment, *Health and Safety at Work etc. Act 1974* (HMSO, 1974)

Control of Asbestos at Work Regulations 1987 (SI 1987 No. 2115) (HMSO, 1987)

Control of Lead at Work Regulations 1980 (SI 1980 No.1248) (HMSO, 1980)

Health and Safety (First Aid) Regulations 1981 (SI 1981 No.917) (HMSO, 1981)

Ionising Radiations Regulations 1985 (SI 1985 No.1333) (HMSO 1985)

Reporting of Injuries, Diseases and Dangerous Occurrences Regulations 1985 (SI 1985 No.967, 1987 Nos 335 & 2112 (HMSO, 1985 & 1987)

Control of Substances Hazardous to Health Regulations 1994 (SI 1994 No.3246) (HMSO), 1995)

Health and Safety Commission, *General COSHH ACOP (Control of Substances Hazardous to Health, Carcinogens ACOP (Control of Carcinogenic Substances) and Biological Agents ACOP (Control of Biological Agents) Control of Substances Hazardous to Health Regulations 1994* (HSE Books, 1995)

Electricity at Work Regulations 1989 (SI 1989 No.635) (HMSO, 1989)

Noise at Work Regulations 1989 (SI 1989 No.1790) (HMSO, 1989)

Personal Protective Equipment at Work Regulations 1992 (SI 1992 No.935) (HMSO, London)

Health and Safety (Display Screen Equipment) Regulations 1992 (SI 1992 No.1043) (HMSO, 1992)

Index